Consolations of Nature

Nature, Culture and Literature

READINGS IN ENVIRONMENTAL HUMANITIES

VOLUME 20

The titles published in this series are listed at *brill.com/ncl*

Consolations of Nature

*Human–Nature Connections in Modern
and Contemporary Anglophone Literature*

Edited by

Šárka Bubíková
Bożena Kucała
Beata Piątek

BRILL

LEIDEN | BOSTON

This book is the result of the research project "Escape into Nature: Consolatory Interactions with the Natural World in Contemporary Literature and Culture" (2023–2024), funded by the programme "Excellence Initiative – Research University" at the Jagiellonian University in Kraków.

Cover illustration: Photograph credited to Marcin Klag.

The Library of Congress Cataloging-in-Publication Data is available online at https://catalog.loc.gov

Typeface for the Latin, Greek, and Cyrillic scripts: "Brill". See and download: brill.com/brill-typeface.

ISSN 1572-4344
ISBN 978-90-04-74410-3 (hardback)
ISBN 978-90-04-74411-0 (e-book)
DOI 10.1163/9789004744110

This book is printed on acid-free paper and produced in a sustainable manner.

Contents

PART 3
Cultivating Nature as a Way of Fostering Communal Justice and Redress

Notes on Contributors

Anton Belenetskyi
is a doctoral student at the Jagiellonian University Doctoral School in the Humanities in Kraków, Poland, where he is working on a Ph.D. thesis concerning uncertainty as the principal experience of the Anthropocene as represented in 21st-century U.S. and Canadian literature. His main academic interests lie in the intersection of ecocriticism, feminist new materialisms, and broadly understood Anthropocene studies. Outside the world of academia and literature, he prefers to spend his free time rearing (and being reared by) his companion species of choice – common houseplants.

Jan Beneš
is Assistant Professor at the University of Ostrava in the Czech Republic, where he teaches survey courses in American and British literature. He received his M.A. from Texas A&M University and his Ph.D. from Masaryk University in Brno, Czech Republic. His research interests revolve around African American literature, culture, and history, and range from Black aviation and aviators of the Harlem Renaissance era to Black environmental literature. He is the co-editor of an edited monograph entitled *Environmental Justice in Ethnic American Literatures* with Lexington Books.

Šárka Bubíková
is Associate Professor at the University of Pardubice, Czech Republic. Her research interests include contemporary American crime fiction, Anglophone children's literature, American ethnic writing and the ethnic Bildungsroman. Apart from numerous articles, she co-authored with Olga Roebuck *The Place It Was Done: Location and Community in Contemporary American and British Crime Fiction* (2023, McFarland) and co-edited *Places and Spaces of Crime in Popular Imagination* (2021). She is the Editor-in-Chief of the journal *American and British Studies Annual* published by the University of Pardubice. She was a Fulbright Scholar at Amherst College, Amherst, MA, and a visiting researcher at UCSB, Santa Barbara, CA. She also writes fiction.

Izabela Curyłło-Klag
teaches in the Institute of English Studies at the Jagiellonian University in Kraków, Poland. Her research interests include the modern British and Irish novel, utopian/dystopian fiction, and the intersections between literature, history and culture. She has published numerous articles on avant-garde and

contemporary writers, and a monograph on representations of violence in early modernist fiction. She has co-edited an anthology of immigrant memoirs, *The British Migrant Experience, 1700–2000* for Palgrave Macmillan, as well as four volumes of critical essays: on literary representations of the past, on dialogic exchanges between literature and the visual arts, on incarnations of material textuality, and – more recently – on representations of housing across media (for Brill/Fink).

Tereza Dědinová

is Assistant Professor at the Faculty of Arts of the Masaryk University in Brno, Czech Republic, where she teaches theory and history of fantastic literature and Czech literature. She has written a monograph (2015) and edited three titles focusing on various aspects of speculative fiction. She has published articles devoted to the fantastic from the cognitive and ecocritical perspective, the representation of the actual world in fantasy, and Czech speculative fiction. Her most recent projects involve co-edited volumes *Images of the Anthropocene in Speculative Fiction: Narrating the Future* (2021) and *Fantasy and Myth in the Anthropocene: Imagining Futures and Dreaming Hope in Literature and Media* (2022).

Aleksandra Kamińska

(Ph.D.) is an assistant lecturer in the Institute of English Studies at the Jagiellonian University in Kraków, Poland. Her academic interests include contemporary British drama, ecocriticism and translation theory. She is also a freelance translator and editor of literary fiction. She has edited, among others, Joanna Bator's *Ucieczka niedźwiedzicy* (2022).

Bożena Kucała

is Associate Professor at the Institute of English Studies, Jagiellonian University in Kraków, Poland, where she teaches nineteenth-century and contemporary English literature. Her research interests focus on contemporary British fiction, especially the historical novel and neo-Victorian fiction. She is the author of *Of What Is Passing: Present-Tense Narration in the Contemporary Historical Novel* (2023), *Intertextual Dialogue with the Victorian Past in the Contemporary Novel* (2012) and co-editor of the book series "Topographies of (Post)Modernity: Studies in 20th and 21st Century Literature in English."

Beata Piątek

is Associate Professor in the Department of Comparative Studies in Literature and Culture at the Institute of English Studies of the Jagiellonian University in

Kraków, Poland. She is the author of *History, Memory, Trauma in Contemporary British and Irish Fiction* (2014) published by Jagiellonian University Press. Her articles in academic journals deal with traumatic memory and commemoration of the Great War in contemporary British and Irish novel, identity and history in contemporary Irish fiction. With colleagues from the Department she organised conferences on *Aftermath: The Fall and the Rise after the Event* (2018) and *Ellipsis: Silence, Absence and Noncommunication in Contemporary Literature* (2019), *Landscape and Identity in British and Irish Literature* (2024).

Olga Roebuck

(Ph.D., M.Litt.) is an assistant professor at the University of Pardubice, Czech Republic. She specializes in cultural identities in contemporary crime fiction and other popular genres. She co-edited several volumes, most recently *Places and Spaces of Crime in Popular Imagination* and co-authored the book *The Place It Was Done*. She is a keen walker and swimmer.

Nataša Tučev

is Associate Professor at the University of Niš, Serbia, Faculty of Philosophy, English Department, where she teaches undergraduate and postgraduate courses on English Modernism, Twentieth Century Anglophone Literature, Literary Theory and Literary Translation. She has published the following books to date: *An Introduction to the Modernist Novel* (2021), *The Secret Sharers: Joseph Conrad's Literary Characters* (2017) and *Inner Emigre: Seamus Heaney's Poetics* (2011). Her academic interests include Modernism, Modernist Novel and Joseph Conrad Studies. She is a member of the Association of Literary Translators of Serbia. Her most notable literary translations include Byron's *Childe Harold* and Coleridge's *The Rime of the Ancient Mariner*.

Ladislav Vít

received his Ph.D. from Charles University, Prague, and now works at the University of Pardubice, Czech Republic. His research interests lie with British interwar writing, literary topography, travel writing and the poetics of place. His major focus is on spatial responsiveness from the perspective of cultural and humanistic geography. His publications include "Feet on the Ground: Landscape in Auden's Late Poetry" (2014), "Poetry and Place in Auden's *Letters from Iceland*" (2016), "Landscape as a Benchmark: Poetics of Place as a Critical Tool in W. H. Auden's Prose" (2018), "Between Blinding and Enlightening: On Auden, Myth and Knowledge" (2022) and "The Ethics of Going North: Moral Geography in Louis MacNeice and Wystan Hugh Auden's *Letters from Iceland*" (2023). In 2021 Routledge published his *The Landscapes of W. H. Auden's*

Interwar Poetry: Roots and Routes, the first book-length study foreground-ing Auden's sense of place as a means for interpreting the work of this cru-cial twentieth-century poet. He is the co-founder and executive editor of the peer-review scholarly journal *American and British Studies Annual,* published since 2008.

Sylwia Janina Wojciechowska
is Associate Professor in the Department of Literature Studies at Ignatianum University in Cracow, Poland. She holds a Polish degree in Classical Philology and a German degree in English and Italian Philology. She is the author of *Re(Visions) of the Pastoral in Selected British and American post-Romantic Fiction* (2017) and *Nost/algia as a Mode of Reflection in the Autobiographical Narratives of Joseph Conrad and Henry James* (2023). As a Conradian scholar, she has been a Board Member of the Polish Joseph Conrad Society and member of the Polish Academy of Arts and Sciences. Her latest contribution is contained in the *Routledge Companion to Joseph Conrad* (2024).

Introduction: Consolations of Nature

Šárka Bubíková, Bożena Kucała and Beata Piątek

In his Introduction to *Literature as Cultural Ecology* (2017), Hubert Zapf highlights the recent surge in ecocritical scholarship, emphasising literature's unique ability to address the complex and multifaceted relationship between culture and nature. Zapf underscores literature's dual role in offering "civilizational critique" and fostering "cultural self-renewal" (4). Concurring with this view, Roman Bartosch invokes Lawrence Buell's assertion in *The Environmental Imagination* (1995) that the current breakdown in humanity's relation to nature is essentially "a crisis of the imagination." Literature, by nurturing and challenging our imagination, can play a crucial role in addressing this crisis (Bartosch 118).

The term "Anthropocene," popularised by atmospheric chemist Paul Crutzen in 2000, marks an era of human-induced planetary transformation (Ellis). Despite its recent rejection by geological sciences (Ellis), the term remains influential across various disciplines, highlighting the political dimensions and social consequences of geological changes (Zalasiewicz et al.). Scholars working within a broad field of environmental humanities are probing the implications of the rapid planetary transformation "for human societies and their ideas of justice, decency and order" (Zalasiewicz et al.), especially as our pursuit of economic growth threatens the planet's habitability and may eventually result in "humanity's potential social and ecological demise" (Scheffran et al. 131). This interdisciplinary approach underscores the urgency of addressing the environmental crisis through multiple lenses, including literature.

Contemporary ecocriticism emerged in response to growing public anxiety over climate change, environmental pollution and biodiversity loss – issues that Scott Slovic describes as crises "wrought by modern civilization" (15–16). This renewed environmental concern mirrors earlier movements during periods of significant change, such as the Romantic response to the anxieties brought about by industrialisation and urbanisation, or the early 20th-century fears of technological progress and the calamity of World War I. Those historical precedents reveal a recurring pattern: during times of rapid and unsettling change, there is a tendency to seek solace and renewal in nature. Today, new anxieties, including the impact of artificial intelligence, ongoing war conflicts, global pandemics and climate change, compound those historical fears, making the role of literature in addressing these issues even more critical. Dale Jamieson, in "The Anthropocene: Love It or Leave It" (2017), articulates

© KONINKLIJKE BRILL BV, LEIDEN, 2026 | DOI:10.1163/9789004744110_002

the unprecedented nature of the current crisis, emphasising the paradox of human power and "the loss of agency" (14–15). He notes that while humanity has significantly altered the planet, there is a simultaneous sense of helplessness in the face of these changes. This paradox is central to understanding the contemporary environmental crisis and underscores the need for imaginative and creative responses.

Melanie Braunecker and Maria Löschnigg, in their Introduction to *Green Matters: Ecocultural Functions of Literature* (2019), argue that literature offers vital perspectives and alternative understandings that science alone cannot provide. They assert that literature's creative potential can serve as "a regenerative cultural element" (3), opening up novel ways of thinking about and engaging with the environment. This perspective aligns with Zapf's view of literature as a means of cultural self-renewal, capable of inspiring both individual and collective action.

In *Twenty-First-Century Fiction: A Critical Introduction* (2013) Peter Boxall observes the prevalence of environmental apocalypse and dystopian visions in recent literature (14); in a similar vein, Astrid Bracke, the author of *Climate Crisis and the 21st-Century British Novel* (2018), argues that "climate crisis has become part of the contemporary literary imagination" (2). However, alongside these bleak portrayals, literature also inspires constructive engagement, social activism, and a renewed appreciation of nature's restorative and uplifting role. As Alexandra Pollard (2020) contends, "writing about the natural world [...] can make it all seem worth fighting for." This duality in literary responses – between alarmist and hopeful narratives – reflects the complexity of the current environmental discourse.

The enforced distance from nature in modern civilisation often triggers a desire to reconnect with and draw solace from the natural world. This impulse is not merely nostalgic but may be seen as a necessary response to the ecological crisis. By accentuating anew the numerous consolations that nature can offer, literature inspires action and hope.

In this collection of essays,[1] we explore how literary works focusing on the hopeful, restorative and consolatory effects of human–nature encounters provide valuable insights and foster a deeper connection with the natural world. However, the essays' authors are fully aware of the potential pitfalls and simplifications of an overly optimistic approach to human–nature entanglements and, while focusing on their more beneficent aspects, they are by no means uncritical. Consolation is not tantamount to a solution to our environmental

---

1   This work was partly supported by the programme "Excellence Initiative – Research University" at the Jagiellonian University in Kraków.

problems, but it can counter despair and apathy in the face of seemingly unstoppable environmental destruction. The following essays demonstrate that literature, through its imaginative and transformative power, plays a vital role in addressing the environmental challenges of our time.

As David James points out in the Introduction to *Discrepant Solace: Contemporary Literature and the Work of Consolation* (2019), the idea of consolation customarily connotes "the prospect of relief, [...] hints of revival and eventual endurance" (1), and traditionally serves as an antidote to "grief, remorse, longing, and dread" (21). In fact, "consolation" may be used in different interrelated contexts. In the 1822 edition of Samuel Johnson's *Dictionary*, "consolation" is treated as largely synonymous with giving "comfort" and the "alleviation of misery." In Johnson's explication, a work of art or a rhetorical piece may be described as "consolatory" if it produces such an effect (James 12). The Latin *consolatio*, according to *Oxford Latin Dictionary*, signifies, first, "the act of consoling or an instance of it," second, "the fact of being consoled," and, third, "a consoling fact or circumstance" (Jedan, Maddrell and Venbrux 3). *The New Oxford Dictionary of English* defines consolation as "the comfort received by a person after a loss or disappointment," or "a person or thing providing such comfort" (392). The present volume, *Consolations of Nature*, comprises articles on selected (predominantly British and American) modern and contemporary fictional and non-fictional texts as well as cultural movements which share the perception of the natural world as a source of comfort and consolation.

The tradition of consolation, which the title of this collection obliquely invokes, originated in ancient Greece and was shaped by the clearly defined practical goal of offering solace and comfort to a person faced with bereavement, old age, illness, destitution, exile or other personal calamities. However, as Antonio Donato contends, the activity of consolation did not consist in offering sympathy to the grieving individual but in persuading the addressee that despair was not an adequate response to their predicament by providing a philosophical perspective on the hardships they suffered, or by recommending practices that could improve one's emotional and psychological condition (398–399). Consolation could also be self-addressed, with the aim of restoring one's inner balance. This was the purpose of Cicero's now lost *Consolatio*, in which the Roman writer tried to come to terms with the loss of his daughter.

The mode of self-consolation was continued in medieval writing, arguably best exemplified by Boethius' renowned *De Consolatione Philosophiae* (*The Consolation of Philosophy* [524 CE]), which features a female personification of philosophy helping her disciple prepare calmly for the prospect of death (Donato 399). Christoph Jedan describes Boethius' work as an iconic text of the pre-modern strand of consolation, which, in embracing both philosophical and theological arguments, is firmly grounded in a moral and metaphysical framework (Jedan 24). *De Consolatione Philosophiae* demonstrates that although the tenor of Christian solace inevitably differed from pagan consolation, there are undeniable continuities between these traditions since certain classical themes and rhetorical modes were espoused by medieval and early modern writers. It was also in antiquity that "a common stock of consolatory topics" (Scourfield qtd. in Tokarski 451) and a canon of consolatory genres, comprising "the letter, the dialogue, the oration, the elegy, and the manual" were established (McClure qtd. in James 22) before these genres were embraced and adapted in subsequent epochs.

Rooted in the ancient ideals of inner peace, calmness and satisfaction (Hadot qtd. in Tokarski 451), consolatory writing customarily consisted in attempts to help the addressee restore or achieve the desired state by reducing or eliminating the "disruptive" emotions of grief and anxiety. Hence the strongly rhetorical aspect of this writing – as Mateusz Tokarski notes, it had much to do with "education, and even to some extent with chastising, admonishing, and reprimanding" (451). The activity of consolation entailed a change of the grieving person's perspective, offering them "a new horizon of thought" and the ability to reconceptualise their situation so that the original cause of distress could be viewed differently and ultimately stripped of its unsettling character (Tokarski 451–452). Essentially, the consolation of philosophy may be defined as "a process of rational reconstruction of the conceptual horizon of the sufferer that is motivated by ethical theories about the good life and grounded in metaphysical theories about the world" (Tokarski 454).

Tokarski observes that the contemporary practice of consolation is primarily located in the field of psychology and is underpinned by a different approach to the problem of grief, anxiety and distress. The widely accepted mode seeks to analyse and heal such emotions rather than deprecate them as inappropriate responses to one's predicament (Tokarski 454). Hence, the consolations of psychology stem from strategies of emotional coping (Tokarski 449; cf. also Kunkel and Dennis). Of course, there is also a popular sense of consolation, involving distraction and the assurance of things becoming better (Tokarski 449). Nonetheless, in his article "Consolations of Environmental Philosophy" (2021) Tokarski probes the possibility of adapting the ancient tradition of philosophical consolation to the contemporary problems of human cohabitation with wildlife. He argues that the philosophical solution of a change of perspective, which in this case involves starting to perceive oneself as a part of an ecological community, may result in greater happiness achieved through the transcendence of an egocentric, human-centred approach to animals. The fear, anxiety or distress experienced in encounters with other creatures may be assuaged or overcome (454–458). At the same time, however, the writer acknowledges the limitations of philosophical arguments in addressing environmental issues, just as ancient authors, such as Cicero, were prepared to doubt the possibility of successful consolation when faced with their own private grief (Tokarski 463). Similarly, the articles in this volume investigate diverse ways of seeking out the consolations of nature while acknowledging the inevitable limitations and impediments of such a response to personal or communal problems.

Despite its culturally variable manifestations, the need for consolation remains constant. Although the traditional genres of consolatory writing have become obsolete, contemporary literature continues to respond to grief, sorrow, trauma and despair, yet the term itself is rarely employed in literary criticism. In *Discrepant Solace*, David James argues that whereas "[e]xplicitly negative states of being – shame, melancholy, fury, despair, trauma itself – are consolation's stormy bedfellows," consolation is now "a rather more illicit citizen of literary studies" (39). In his book, James discusses several late twentieth- and twenty-first-century narratives which, according to the author, exemplify the enduring presence of consolation in contemporary writing (34). James, however, takes issue with what he sees as the naïve and complacent meanings of consolation, understood as the act of offering comfort, solace or advice on how to deal with a personal crisis. Instead, he emphasises the ethical, ideological as well as formal complexity of the chosen narratives that deal with grief, sorrow and distress. James argues that they perform the work of consolation indirectly, by representing the experience of crisis and finding an adequate form for such a representation.

The volume *Consolationscapes in the Face of Loss: Grief and Consolation in Space and Time*, edited by Christoph Jedan, Avril Maddrell and Eric Venbrux (2018), has a much broader scope than James's *Discrepant Solace*, but shares with it an attempt to enhance the utility of consolation "as a concept and analytical tool" in contemporary scholarship (4). As the editors claim in the Introduction, the notion of "consolation" has been largely neglected to date, and if it is referenced at all, this is usually done in an intermittent and unsystematic manner, more or less interchangeably with "comfort," "coping," "resilience" and "solace" (2). Their interdisciplinary appraisal of consolation comprises contributions from the fields of geography, sociology, anthropology, history, philosophy and religious studies. This diversity of approaches is intended to instantiate the viability of consolation and its relevance in modern thought. Accordingly, in the chapter "What Is Consolation: Towards a New Conceptual Framework" Christoph Jedan argues that while the term may have become outmoded or suspect, consolatory practices have never actually disappeared but have taken on novel forms in changing historical and cultural circumstances (20–21). Thus, for instance, the twentieth century saw the emergence of "professionalised consolation" which replaced the earlier religious and pre-scientific connotations of the discourse of consolation with the more acceptable ideas of "coping" and "resilience" (30). Nonetheless, Jedan contends that the new strand still represents a continuation of the consolatory tradition (31). Making a case for the comprehensiveness and inclusiveness of the concept, he distinguishes the defining characteristics of consolatory writings and practices, such as the goals of attenuating grief, reducing distress, increasing resilience and restoring "normal functioning" (33, 35). It is in this inclusive sense that the term "consolation" is employed in the title of the present volume.

The recent rise of ecopsychology and ecotherapeutic methods, which is one of the manifestations of enhanced contemporary ecological awareness, exemplifies and implicitly confirms Jedan's thesis concerning the continuity and professionalisation of former consolatory practices. On the other hand, however, ecopsychology also highlights the role of nature as a potentially powerful source of consolation.

While the idea that natural landscapes and elements (such as fresh air, sunshine) can promote human wellbeing has deep roots, it took a more specific shape in the concept of the nature cure in the nineteenth century in response to the pressures of industrialisation and urbanisation. Inspired by the principles of the nature cure and related health movements, more structured therapeutic practices connected with outdoor activities, so-called wilderness therapy, evolved in the mid-twentieth century. In the seminal *Wilderness and the American Mind* ([1967] 2014) Roderick Frazier Nash noted that "'wilderness therapy' came to have increasing prominence in mental health literature of the 1970s" (266). Both the nature cure and wilderness therapy originated as anthropocentric concepts, focused on human needs and experiences, and implicitly based on the dichotomy between nature and culture. Further, as ecotherapist Laura Marques Brown (2020), founder of the Decolonizing Wilderness Therapy programmes, argues, they were attuned to healing people representing the dominant culture, disregarding (unintentionally even harming) non-dominant participants. While ascribing value to nature on the one hand, seeking a cure in nature can also lead to the ironic result of environmental degradation on the other. Nash poignantly sums up that the century-long struggle to ignite a love for wild nature may, in the end, cause nature to be loved to death due to overexploitation (316). Samantha Walton, in *Everybody Needs Beauty: In Search of the Nature Cure* (2021), expresses similar reservations about the potentially negative environmental impact of promoting nature cure tourism and encourages readers to consider more sustainable and equitable ways to integrate the healing propensity of nature into their lives. Thus, there is currently a discernible shift towards more holistic and ecocentric approaches that also strive to overcome the nature–culture dichotomy.

Ecopsychology aims to transcend this dichotomy by fostering a deeper, more integrated relationship between humans and the natural world and encourage sustainable interactions with the environment. Ecopsychologists treat the threat to the ecosystem as symptomatic of culture's "dissociation" from the whole of which it is a part, and this, in turn, is recognised as one of the causes of the current crisis. Hence the proposed strategy of dealing with it is to reverse the process and appreciate the "therapeutic value" of bringing the human and the other-than-human together (Totton and Rust xvii–xviii). In their Introduction to *Vital Signs: Psychological Responses to Ecological Crisis* (2018), Nick Totton and Mary-Jayne Rust claim that "a sane culture" must be grounded in ecological consciousness (xviii). The projects, therapies and case studies described by the contributors to their volume stress the need for a holistic approach and for building relationships with the natural as an impulse for creative transformations. Kelvin Hall, among others, argues that buried deep in our psyche is "a spectrum of intimate connection" with other species, landscape and the elements, which should be uncovered during psychotherapy (79). Empirical findings suggest that experiencing a wider sense of belonging not only leads to personal healing, but is likely to result in greater commitment to environmental causes (Crompton 204). Should the present ecological crisis grow even more serious, the future role of ecopsychology, according to Totton and Rust, will probably be "to help people manage the pain and despair that will accompany 'the end of the world', and to preserve some sort of hope" (xviii).

Comparable premises regarding the healing and restorative potential of nature have been adopted by Rebecca Crowther, a transdisciplinary ethnographic researcher, in the projects she expounds on in her book *Wellbeing and Self-Transformation in Natural Landscapes* (2019). Crowther invokes three interconnected theories that try to account for the causal link between human interaction with nature and a sense of wellbeing. The most comprehensive of these theories, "the biophilia hypothesis" (originally put forward by Edward O. Wilson) proposes that "life and lifelike forms" appeal affectively to our inborn need to seek connections with other living beings. At the core of the "stress-reduction theory" is the claim that the soothing effect of exposure to certain natural environments is an evolutionary development which ensures wellbeing and survival. Finally, the "attention restoration theory" emphasises the possibility of revival of energy and mental abilities in contact with nature by allowing our senses to relax and our "spontaneous attention" to be reinforced (cf. Crowther 18–19). Deliberately countering the currently widespread preoccupation with human estrangement from nature and with the ensuing environmental crisis (24–25), as part of her projects Crowther took her study groups on several excursions across rural landscapes of Scotland in order to observe the impact of non-urban places on individuals as well as on the relations within the groups. In her conclusions, the author claims that the exercise led to positive personal and interpersonal transformations, which she is inclined to trace to innate ethical impulses, supposedly bolstered by exposure to the realm of nature (277). While the initial motivations for journeying to natural spaces varied from seeking opportunities for recuperation, recovery, lifestyle change or mental transformation (4), the shared outcome declared by the participants consisted in experiencing "'something' positive" (293) in encounters with the natural world.

Indeed, these and countless comparable ventures and projects implicitly or explicitly evoke Edward O. Wilson's notion of "biophilia," defined by him as "the innate tendency to focus on life and lifelike processes" (1) in his much-debated 1984 book *Biophilia: The Human Bond with Other Species*. Drawing on scientific evidence, the author ascribes biophilia to our evolutionary conditioning and argues that the biologically motivated, self-interested need to preserve life in its diverse forms could be rationalised to form the basis of a conservation ethic (139–140). In his 2013 article "Biophilia and the Conservation Ethic" Wilson affirmed his biophilia hypothesis, rewording it as "the innately emotional affiliation of human beings to other living organisms" (31). In the words of Stephen R. Kellert, the biophilia notion "powerfully asserts that much of the human search for a coherent and fulfilling existence is intimately dependent upon our relationship to nature" (43). Wilson's idea may be regarded as an attempt to give a modern scientific account of what is a widespread and age-old intuitive recognition of the beneficent effect that nature may have on us. As Samantha Walton says, in this instance science essentially confirms "things that have long felt natural and commonsense" (4). Walton took part in a similar exercise as the one described by Crowther. Recounting her experience, she stresses the extent to which concepts such as Wilson's have taken hold of the popular imagination: "Whether they're quoting scientists or not, most of the people gathering in the wood believe in Wilson's biophilia. We are natural animals" (11).

Yet, without discounting this view, Walton is clearly mistrustful of an uncritical pursuit of "the nature cure," as suggested earlier. For one thing, "the nature cure" should not be taken literally – there are conditions for which we need to turn to modern medicine rather than nature (8). Also, there are the risks of naïve optimism and simplification, such as the glib exploitation of the revival of "the nature cure" idea (5) to the negligence of actual commitment in the face of the calamities we face today: "a crisis of mental health, social injustice and environmental devastation," which, according to Walton, are "inextricably entwined" (14). Therefore, it is in the growing awareness of the interdependencies between the health of nature, the health of society and the health of the individual that Walton also detects some grounds for a hopeful outlook, provided, however, that the notion of "the nature cure" is treated holistically (14).

This volume examines a variety of motivations for engaging with nature and employs the term "consolations" for the positive outcomes of such individual and communal engagements. The reflections on consolation proposed by the German scholar Hans Theo Weyhofen in 1983 are especially pertinent here. Although they clearly hark back to the ancient philosophical model, they also serendipitously resonate with much of the contemporary ecological thought:

> Consolation's point of departure is a difference – more specifically, a contradiction – between the human and the world. On one side we find the interests, wishes and goals of the human being; on the other side there is the world, which does not comply with those interests, wishes and goals. Consolation is the answer to the suffering that is caused by that difference. The goal of consolation is to remove the difference and to produce a reconciling identity. (qtd. in Jedan 19, trans. by Jedan)

More recently, Kirsten Anne Tornøe et al. (2015) formulated a similar framework for consolation: "consolation is needed when a human being feels alienated from him or herself, from other people, from the world and from his or her ultimate source of meaning."

From the perspective of environmental humanities, the pursuit of the consolations of nature is underpinned by the desire to overcome the human–nature dualism which, as many commentators have contended, "has shaped modern civilisation" (cf. Oppermann 39), and replace it with a notion of "co-presence and interdependence" (Iovino qtd. in Oppermann 45). The idea of "the interconnectedness of all living and non-living things" (Morton 28) is encapsulated, among others, in Timothy Morton's well-known concept of the mesh, put forward in his book *The Ecological Thought* (2010). Morton's notion of all-encompassing multidirectional connectiveness entails the renunciation of a hierarchy of beings, including the idea of human superiority. According to Morton, it is the ecological crisis that made us aware of this universal interdependence (30). These ideas are central to his more recent book *Dark Ecology: For a Logic of Future Coexistence* (2016), in which he denounces the anthropocentric logic of the past twelve thousand years and champions the need for a new ecological awareness based on a nonanthropocentric logic of coexistence.

"Dark ecology," a notion with which Morton "puts hesitation, uncertainty, irony, and thoughtfulness back into ecological thinking" (*Ecological Thought* 16) in turn provides inspiration for Heather I. Sullivan, who proposes her own ecocritical trope – the dark pastoral. This trope is a hybrid of "'dark' (ironic, posthuman, postmodern, polluted) and 'green' (the sentimental and 'artificial natural' of the pastoral that is also biophilic)" (Sullivan 20–21). Closely related to Terry Gifford's concept of the post-pastoral (1999), which similarly challenges the romanticised view of nature and emphasises the interconnectedness of humans and the natural world while promoting humility and respect towards the environment, Sullivan's dark pastoral is adapted to "the troubling catastrophe-centred scenarios so popular in the fossil-fueled era of the Anthropocene" and, as she says, can be used for "exposing the dynamics of power and agency in relation to material nature-culture" (19–20). By problematising the traditional connotations of pastoralism, Sullivan challenges attempts to return to a pristine state of nature as "naively retro-nostalgic" (19).

Whether naïve or not, turning to nature in search of comfort and consolation is a fairly common response to personal or collective anxieties. In *Losing Eden: Why Our Minds Need the Wild* (2020) the journalist and nature writer Lucy Jones observes that our self-induced severance from the "crucial bond" with nature has produced the opposite tendency to restore it. Jones highlights the renewed need for cultivating this vital connection in the context of the recent ecological crisis: "It's in the last five years that the consequences of our separation from nature are really being shown to us – through species loss, biodiversity loss, the climate crisis, but also the health repercussions for humans. The more time goes on, the more we see how crucial that link can be, and how nature can help all manner of mental health problems" (qtd. in Pollard).

Yet, the appreciation of nature and a search for a way of relating to it is a phenomenon with a long tradition behind it, although, admittedly, it has recurred in different guises and on different scales, with a concomitant evolution of the human perspective on nature. Morton himself speculates that Henry David Thoreau's once isolated and idiosyncratic experience of amazement at the unity of all life is now understood by all (*Ecological Thought* 33). Whereas it is debatable where the origins of modern-day environmentalism should be sought, there are solid grounds for tracing its inspirations to the Romantic quest for the sublime wilderness, contemplative solitude, the wish to escape from the pressures of modern life and the discovery of forms of co-existence with other species. As Kate Rigby points out in her monograph *Reclaiming Romanticism* (2021), the notion of human entanglements with other living beings was first propounded by Fredrich Schlegel, who coined the term "sympoiesis" to describe it (Rigby 18). Whereas in the North American context it is Henry David Thoreau who is generally credited with inspiring "environmental imagination," in Britain this role is attributed primarily to William Wordsworth. Rigby avers that his solitary reveries in nature were, on the one hand, related to the tradition of contemplative practices and a pantheistic version of Christian spirituality, and, on the other hand, they paved the way for modern, secular ecopoetics (28–29). Although, as she argues, the poet's sense of the sacred drive within all things and their dynamic interrelations was specifically indebted to Spinoza (36), she regards Wordsworth as one of the forerunners of the modern idea of the interconnectedness of human and other-than-human entities. Likewise, in their article "Engaging the Imagination: 'New Nature Writing', Collective Politics and the Environmental Crisis" (2018) the ecocritics Kate Oakley, Jonathan Ward and Ian Christie trace the antecedents of contemporary engagements with nature as far back as late eighteenth- and early nineteenth-century responses to the encroachment of urbanisation, industrialisation and the gradual loss of countryside and wilderness. They observe similar continuities of "mourning, celebration, assertion of an aesthetics and ethics of land and the wild, coupled with advocacy of conservation" in Britain and the USA throughout the nineteenth and twentieth centuries, leading to the present-day resurgence of various forms of writing devoted to nature (689–690). In the first decades of the twentieth century, the idea of a return to nature was fuelled by a sense of civilisational crisis brought on by the advance of urbanity, technology as well as the devastating experience of the Great War. Now, in the first decades of the twenty-first century, the tendency to reconnect with the natural seems all-pervasive. Oakley, Ward and Christie observe a proliferation of popular writing which gives advice on how to return to nature:

> Visit a bookshop, even in the most urban of locations, and there will be a table groaning with books about the countryside. Moving to the countryside, moving *back* to the countryside, the loss of countryside and the rewilding of the countryside, farming, falconry, dry stone walling and above all, walking. (688)

To take a transhistorical and transcultural perspective, the (re)turn to nature fulfils certain personal needs, which may be both bodily and mental, but occasionally becomes a collective impulse when a major civilisational exigency arouses escapism, a desire for contact with an unspoilt, pristine or rural environment, or a yearning for restorative connections with other forms of life. Responsible stewardship of nature, such as land cultivation or nature conservation, may also become a way of redressing social wrongs, or rectifying errors in technological advancement. Thus, what Capaldi et al. (2) describe as "the age-old belief" that "connecting with nature" is beneficial for individual wellbeing may be explored in numerous contexts, both including and transcending personal experience (cf. Crowther 21). Of course, the forms that such tendencies take depend greatly on how "nature" is construed in a given epoch. Responses span a wide spectrum, from idealisation, pastoralism, a search for "the egoistical sublime" or for a refreshing, primal antidote to civilisational discontents, to scepticism about the possibility of return to nature, and to scientifically- or ecologically-oriented stances. In her overview of British nature writing at the turn of the twenty-first century, Deborah Lilley detects a strand of works that "move beyond a binary understanding of 'natural landscapes' to explore the intersections of people and place and, in doing so, begin to re-envision the concepts of 'nature' and 'wildness,'" in contrast to the solitary contemplations of Romanticism [3]. This writing, according to Lilley, is new not only due to its peculiar blend of "autobiography, travelogue, natural history, and popular science," but also thanks to its enhanced awareness of the environmental crisis, which necessitates a redefinition of how we relate to nature (Lilley [4]). Prominent in Lilley's discussion is the claim that, given the extent of human impact on natural processes, the nature–culture binarism is becoming increasingly blurred, which, in turn, is reflected in numerous instances of contemporary nature writing.

Nowadays there is also a growing recognition that in seeking to benefit from the natural world we cannot objectify it or treat it as a cultural construct but rather acknowledge our position as an integral part of the ecosystem. In discussing the "values" of nature, the environmental scientists William T. Borrie and Christopher A. Armatas (2022) distinguish broadly between instrumental and intrinsic values. The former presuppose a view of nature as "an external object" which is assessed in terms of the (educational, recreational, commercial, etc.) utility it provides (15, 18). The latter kind of values is underpinned by a recognition that nature is "valuable and important in, and of, itself," independently of its supposed benefits to humans (17). Yet there is also a third type, namely the "relational values" of nature that arise in the course of human–nature relationships. Following Chan et al. (2016, 2018), Borrie and Armatas claim that

> interacting with nature connects one to land, strengthens traditions and encourages contemplation, thus sustaining the relationship between human well-being and nature. It can be said that people belong to a place and must behave virtuously – with relational behaviour such as reciprocity, care, custodianship, or stewardship of places celebrated as duties and responsibilities. (18–19)

Indeed, prominent in contemporary environmental humanities is an understanding of the physical world as a dynamic, relational process (Oppermann 42). Serpil Oppermann argues that the recognition of "relational ontologies" should inform ecocriticism if it is to function as "an effective intellectual response to ecological and political urgencies" (44).

• • •

Whilst the articles that comprise the present volume have little in common with the antiquated genre of consolatory writing, they still diverge from the prevalent ongoing discourse of ecological and civilisational crisis by discussing texts as well as socio-cultural projects that adopt a more optimistic and hopeful approach to the human entanglements with the natural, concentrating on the variegated consolations that such entanglements may yield. The chapters analyse a variety of beneficent human–nature interactions, ranging from nostalgia, escapism, anthropomorphic fantasies, celebrations of the wild coupled with environmental concerns, personal testimonies, to literary fictions about gardening, conservation, and environmental justice. The collection comprises early twentieth- and early-twenty-first century instances of the consolations of nature, predominantly in the British and American contexts. This selection of twentieth- and twenty-first-century cultural and literary engagements with the natural world both testifies to the current yearning for a harmonious, healing bond with the natural world and points to the enduring and recurrent aspect of this tendency. Thus, the volume *Consolations of Nature: Human–Nature Connections in Modern and Contemporary Anglophone Literature* aims to draw attention to an approach in ecocriticism that is often overshadowed by pessimistic or downright apocalyptic accounts.

Despite the dominance of these views, there exists a stream of optimistic and hopeful perspectives within ecocritical literature (see for example Joseph Meeker 1974; Mitchell Thomashow 1995, 2001; John Felstiner 2009; Mark S. Cladis 2020), which foreground the aspects of hope and resilience and focus on the ability of literature to motivate and empower readers. This book continues in this stream and offers a fresh and refreshing perspective on recent challenges to the traditional nature–culture dualism by inspiring hope rather than highlighting a sense of crisis. Another contribution that this collection makes to ecocritical readings is to indicate three possible contexts in which consolatory, restorative or uplifting relations with nature may be explored further in modern and contemporary literature and culture.

The volume consists of eleven chapters grouped into three parts: Part 1: "(Re)Connecting with Nature as a Response to Modernity," Part 2: "Elemental

Entanglements as a Source of Personal Consolation," and Part 3: "Cultivating Nature as a Way of Fostering Communal Justice and Redress."

The point of departure in Part 1: "(Re)Connecting with Nature as a Response to Modernity" is the observation shared by many early twentieth-century intellectuals that the condition of modern man is a state of rupture and divide between the human world and the environment. The writers, artists and activists whose work is discussed in this section produced a wide variety of responses to what Stephen Spender in *The Struggle of the Modern* (1963) called "a screen set up between nature and man." They ranged from optimism regarding the human potential to return to nature for the sake of re-establishing the lost rapport and connectedness, through pragmatic and consolatory approaches, to pessimism insisting on the existence of an irremovable divide between the two entities. In the opening chapter, "*Consolatio naturae* in Joseph Conrad's *The Mirror of the Sea*," Sylwia Janina Wojciechowska reassesses Conrad's meditation on the broken connection with nature as experienced on the cusp of the twentieth century. Focusing on *The Mirror of the Sea* (1906), the author argues that this autobiographical narrative is a testimony communicating the redefined status of man in the world of nature. As Wojciechowska demonstrates, the consoling potential of Conrad's maritime chronicle lies in the set of images that represent the human being as ultimately belonging within the natural world rather than with machines and technology. In the next chapter, "Nature, Language and Silence in D. H. Lawrence's *St. Mawr*," Nataša Tučev begins with the writer's quest for alternative forms of communication in the light of his frustration with inadequacies of human language. She demonstrates how Lawrence uses metaphor and literary synaesthesia in order to give voice to natural entities. Drawing on the writings of Christopher Manes and David Abram, Tučev contends that Lawrence's endeavour to "restore relations with the natural world as a cure for the alienated human condition" places his texts in line with contemporary ecocritical concerns. The first two chapters demonstrate that whereas both Conrad and Lawrence draw on their autobiographical experiences to reflect on the consolatory powers of nature, they also reflect some of the anxieties of the modern age.

The next two chapters in this part scrutinise popular movements promoting a return to nature in interwar Britain. Ladislav Vít's chapter entitled "Crossing the Rift: Interwar Intellectuals and Mass Escapism to Nature" examines a broad range of reactions to the widespread escapism to the countryside. Vít discusses the development of the concept of the countryside as a tool for fostering individual, mental and physical wellbeing, citizenship, and a more refined nation. He goes on to examine critical reactions to the open-air ethos in the writings of Henry Scott Holland, George Macaulay Trevelyan and Wystan Hugh

Auden. Set within the same period and cultural context, the fourth chapter of the section is devoted to the analysis of the initial, apolitical phase of John Hargrave's Kibbo Kift Kindred, which Izabela Curyłło-Klag, the author of "Reclaiming Nature: John Hargrave's Kibbo Kift and the Vision of Post-World War I Renewal," sees as a unique creative culture, rooted in an appreciation of the natural world. As an antidote to the civilisational crisis brought on by the Great War, Hargrave proposed a distinctive programme of moral and physical regeneration. The chapter challenges the recent revisionary trend in modernist studies, which often portrays the modernist movement as endorsing the destructive facets of modern civilisation. The essays in Part 1 show that the early twentieth-century intellectuals made a significant contribution to the discussion concerning the viability of a re-connection to the world of nature by bringing into focus some of the issues that are still relevant for the contemporary environmentalist debates.

This part closes with Tereza Dědinová's chapter "Beyond Dualism: Human–Nature Connections as Anthropomorphic Interaction in Contemporary Environmental Literature," which builds a bridge between the reflections on the relationship between man and nature as discussed in the writings of Conrad and Lawrence and the twenty-first-century problems of the Anthropocene as presented in contemporary fiction. Dědinová opens the theoretical part of her chapter with a discussion of the philosophical origins of the culture–nature dichotomy to demonstrate how the mechanistic materialism of Hobbes and Descartes resulted in an alienation of human beings from nature and how, in consequence, we view ourselves not only as separate from but also opposed to nature. Referring to Timothy Morton's concept of the mesh, i.e. interconnectedness of all living and non-living things, the author of the chapter proposes a reading of selected works by Paolo Bacigalupi, Richard Powers and Terry Pratchett in which the anthropomorphism of nature is used to subvert the culture–nature dualism. She demonstrates that even in the apocalyptic visions of the future of our planet writers create characters who still derive comfort, consolation and a sense of meaning from interactions with nature.

Part 2: "Elemental Entanglements as a Source of Personal Consolation" engages with transgressions of the nature–culture dualism through individual experience of multivarious interactions with other beings and natural entities. Its constituent chapters trace intensely subjective, embodied responses to the environment understood as a living, thriving mesh. In the analysed texts, human encounters with the more-than-human grow out of subjective, physical, multisensory experiences. The protagonists immerse themselves in the realm of nature in a variety of ways, which allows them to (re)discover complex forms of interconnectedness as well as transcorporeality.

The more-than-human world is presented as not only a form of external reality but also as shared participatory presence. By enlarging their perspective and reconfiguring it to encompass individual complex entanglements in the environmental network, the narrators or protagonists achieve an enhanced sense of self, remapping the boundaries of their selfhood in a way which enables them to participate in the continuing flux of life. Thereby, in narratives which recount individual encounters with the more-than-human, consolation, sustenance and the possibility of renewal are discovered through the newly achieved sense of belonging. The human–natural interconnectedness described in particular texts is an analogue to their formal and generic diversity, which encompasses fiction, autobiographical elements, semi-scientific discourse and new nature writing.

The first two chapters of this part discuss representations of elemental entanglements in hybrid texts which typify contemporary British new nature writing, recounting their authors' restorative and healing connections with the natural world. Both begin by reflecting on the genre of new nature writing, its definitions and characteristics. In the opening chapter, "Soil Solace in New Nature Writing," Bożena Kucała analyses *The Grassling* (2019) by the British-Kenyan writer, poet and environmental activist Elizabeth-Jane Burnett. Burnett's strongly autobiographical account blends factual and poetic modes, describing the environs of her English family's native village in Devon. The writer's deeply personal engagement with the place, its landscape, wildlife and in particular the soil that sustains all life, ensuring its continuity and renewal, were prompted by her father's illness. Faced with his imminent passing and confronted with the larger problem of human mortality, Burnett seeks consolation in nature. Approaching human life and culture as intertwined with the natural world, rather than distinct from it, enables her to attain a perspective from which it is possible to reconfigure an individual death as a stage in the enduring cycle of life. The attempts at physical and sensory merging with nature correspond to those passages in the book in which the author invents a language for narrating her imaginary metamorphosis into a blade of grass, to become something literally rooted in the soil. Transcending the limits of human individuality and acknowledging one's part in the living mesh lead to personal consolation and inner balance.

The transcendental dimension of encounters between the human and the natural is also the subject of Olga Roebuck's discussion of another two texts representative of British new nature writing: Roger Deakin's *Waterlog* (1999) and Hugh Thomson's *The Green Road into the Trees* (2012). In "More Than Backpacks and Swims: British Nature Writing and the Consolatory Role of Nature" Roebuck emphasises Deakin's and Thomson's focus on the uplifting and enriching experience of encounters with nature. As she points out, the connection is intensely physical and literal in Deakin's *Waterlog* – the author swims in different locations around Britain. Both Deakin's immersion in the watery element and Thomson's walks around rural England provide them with opportunities for entering into meaningful relationships with particular places. Deakin's close engagement with the natural environment also triggers reflections on ecological issues, questions of land and water ownership, stewardship and conservation. Hence, as Roebuck argues, the author's personal reconnections with nature cannot be described as self-indulgent escapism. Similarly, Thomson's wanderings are motivated by the desire to build connections with rural places and their inhabitants rather than escaping from contemporary problems. Like Deakin, Thomson treats his experiences in the English countryside as a springboard for observations on the human–nature interactions in an age when no pristine environment can be said to exist. Roebuck demonstrates that, irrespective of their shared ecological concerns, each author values his subjective and physical embeddedness in nature. Instances of such re-integration and the concomitant sense of being absorbed into the elemental and regenerative cycle of life lead to an expansion and transformation of the self, enabling individuals to derive comfort and consolation from the harmony between themselves and the environment.

The healing potential of personal encounters with the watery element is explored in the two texts analysed by Aleksandra Kamińska in the final chapter in this part, "'Making Things Right': Fostering Relationships of Care in Narratives of Feminist Hydrocommons." Both the non-fiction work *Braiding Sweetgrass* (2013) by Robin Wall Kimmerer, a Native American writer and scientist, and "Tikkun Olam" (2022), short fiction by the Polish writer Joanna Bator, explore women's restorative connections with water. The theoretical background to Kamińska's comparative analysis is based on recent studies of water as gendered materiality, such as Astrida Neimanis's claim in *Bodies of Water* (2017) that water links female bodies "to other bodies, to other worlds beyond our human selves," which results in the creation of "a more-than-human hydrocommons" (2). Both narratives analysed in this chapter recount women's experience of reclaiming a body of water, and, in the process, regaining control of their lives and building a renewed sense of self. The numerous parallels between the two texts show how the cultivation of nature enables the female protagonists to immerse themselves in the more-than-human both physically and mentally. Consequently, the human–nature encounters in *Braiding Sweetgrass* and "Tikkun Olam" may be said to have tangible consolatory, liberating and transformative effects.

Part 3 is called "Cultivating Nature as a Way of Fostering Communal Justice and Redress." While sharing the previous section's approach to nature as a primarily relational notion, this section stems not from the interest in more-than-human entanglements per se, but rather from the attentiveness to the irreducible differences that continue to underwrite the broadly understood human–nature interdependency. In order to examine such differences more thoroughly, the section situates itself in the current debate around the idea of environmental stewardship. Seen as a potentially effective tool in promoting improved human–environment interactions (Bennett et al. 597), environmental stewardship pertains to cultivating a more mindful and responsible attitude towards land and the natural world in general. The section's chapters engage with variegated narratives of (re)imagining, proposing, or practising stewardship and its relational ethics not only as humans' responsibility towards nature but also as an affirmative reciprocity between the two. They view the notion of stewardship from indigenous, black, and feminist perspectives as an ethical relationship to land, wildlife, and community.

Reflecting the expertise of the chapters' authors, the analysed texts come mainly from British and American literature. They cover a variety of literary genres (novel, poetry, crime fiction, essay, memoir), emphasising that literary renderings of environmental stewardship issues are not limited to any particular literary tradition or form of expression. The settings are multispatial, including urbanscape, garden, farm, and national park. Out of the critical engagement with the reciprocal ethics of stewardship, the narratives (or their protagonists) express hope for a qualitatively better sense of community that extends beyond the confines of the merely human as well as meditate on what constitutes an environmentally sustainable community in the first place. Ultimately, the section intends to contribute to the ongoing imaginative search for how to meaningfully cohabitate with the natural world that we cannot – and should not – disentangle ourselves from.

Turning from the element of water addressed in the previous chapter to the element of soil, Anton Belenetskyi's article "Gardening Forking Paths: The Figure of the Rambunctious Garden and Relational Ethics of Attentiveness and Care" considers the human–nature interrelatedness reflected in a variety of twenty-first-century American literary texts situated in the inherently fluctuating figure of a garden, a unique space hovering between nature and culture, tameness and wildness, design and randomness. The author argues that the act of gardening, whether literal or metaphorical, leads to relational ethics and may help us reconcile with the uncertainties of the Anthropocene. Moving from a garden to a field, Jan Beneš's chapter "'Relearning the Lessons of Land Reverence': Land Stewardship as Environmental Justice in the Writings of Leah Penniman and Natalie Baszile" analyses narratives and memoirs about Black farmers and farming communities, namely Leah Penniman's *Farming While Black* (2018) and *Black Earth Wisdom* (2023), Natalie Baszile's

Queen Sugar (2014) and *We Are Each Other's Harvest* (2021), and focuses on how they address land stewardship as well as the ethos of Black agrarianism. By addressing African Americans' historical trauma of being enslaved, dispossessed and disenfranchised, as well as by rewriting harmful stereotypes about working the land, the texts also perform acts of environmental justice. Thus the analysis demonstrates how closely environmental stewardship is connected to environmental justice and resistance to white supremacy. The Part's closing chapter by Šárka Bubíková, "Crimes, Wolves and Consolation: National Parks in Contemporary Crime Fiction," centres on two crime novels – Nevada Barr's *Winter Study* (2008) and Charlotte McConaghy's *Once There Were Wolves* (2021) – set in national parks and involving human entanglements with wolves. It shows that even genre literature can provide meaningful meditations on the human relationship to the more-than-human world by incorporating issues such as wilderness therapy, consolation of nature, wildness and nature conservation into crime narratives. Both novels advocate for communal stewardship and draw on the contrast between human (specifically domestic) violence as communally destructive, and animal predation as environmentally sustaining.

Bibliography

Bartosch, Roman. "Literary Quality and the Ethics of Reading: Some Thoughts on Literary Evolution and the Fiction of Margaret Atwood, Ilija Trojanow, and Ian McEwan." *Literature, Ecology, Ethics: Recent Trends in Ecocriticism*, edited by Timo Müller and Michael Sauter, Universitätsverlag Winter Heidelberg, 2012, pp. 95–128.

Bennett, Nathan J., et al. "Environmental Stewardship: A Conceptual Overview and Analytical Framework." *Environmental Management*, vol. 61, 2018, pp. 597–614, https://doi.org/10.1007/s00267-017-0993-2.

Borrie, William T., and Christopher A. Armatas. "Environmental Values and Nature's Contributions to People: Towards Methodological Pluralism in Evaluation of Sustainable Ecosystem Services." *Human–Nature Interactions: Exploring Nature's Values Across Landscapes*, edited by Ieva Misiune, Daniel Depellegrin and Lukas Egarter Vigl, Springer 2022, pp. 13–23.

Boxall, Peter. *Twenty-First-Century Fiction: A Critical Introduction.* Cambridge University Press, 2013.

Bracke, Astrid. *Climate Crisis and the 21st-Century British Novel.* Bloomsbury Academic, 2018.

Braunecker, Melanie, and Maria Löschnigg. Introduction to the Volume. *Green Matters: Ecocultural Functions of Literature.* Brill, 2019, pp. 3–16.

Brown, Laura Marques. "Decolonizing Wilderness Therapy." 16 Oct. 2020, https://anchoredhopetherapy.com/decolonizing-wilderness-therapy/. Accessed 3 Feb. 2025.

Capaldi, C. A., et al. "Flourishing in Nature: A Review of the Benefits of Connecting with Nature and Its Application as a Wellbeing Intervention." *International Journal of Wellbeing*, vol. 5, no. 4, 2015, pp. 1–16.

Cladis, Mark S. "Du Bois and Dark, Wild Hope in an Age of Environmental and Political Catastrophe." *Ecozon@*, vol. 11, no. 2, 2020, pp. 216–223, https://doi.org/10.37536/ECOZONA.2020.11.2.3500. Accessed 24 March 2025.

Crompton, Tom. "Back to Nature, Then Back to the Office." *Vital Signs: Psychological Responses to Ecological Crisis*, edited by Mary-Jayne Rust and Nick Totton, Routledge, Taylor & Francis Group, 2018, pp. 201–210.

Crowther, Rebecca. *Wellbeing and Self-Transformation in Natural Landscapes*. Palgrave Macmillan, 2019.

Donato, Antonio. "Self-Examination and Consolation in Boethius' *Consolation of Philosophy*." *Classical World*, vol. 106, no. 3, Spring 2013, pp. 397–430.

Ellis, Erle C. "The Anthropocene Is Not an Epoch – But the Age of Humans Is Most Definitely Underway." *The Conversation*, 5 March 2024, https://theconversation.com/the-anthropocene-is-not-an-epoch-but-the-age-of-humans-is-most-definitely-underway-224495. Accessed 5 Feb. 2025.

Felstiner, John. *Can Poetry Save the Earth? A Field Guide to Nature Poems*. Yale University Press, 2009.

Gifford, Terry. *Pastoral*. 1999. Routledge, 2nd ed., 2019.

Hall, Kelvin. "Remembering the Forgotten Tongue." *Vital Signs: Psychological Responses to Ecological Crisis*, edited by Mary-Jayne Rust and Nick Totton, Routledge, Taylor & Francis Group, 2018, pp. 79–88.

James, David. *Discrepant Solace: Contemporary Literature and the Work of Consolation*. Oxford University Press, 2019.

Jamieson, Dale. "The Anthropocene: Love It or Leave It." *The Routledge Companion to the Environmental Humanities*, edited by Ursula Heise, Jon Christensen and Michelle Niemann, Routledge, 2017, pp. 13–20.

Jedan, Christoph. "What Is Consolation? Towards a New Conceptual Framework." *Consolationscapes in the Face of Loss: Grief and Consolation in Space and Time*, edited by Christoph Jedan, Avril Maddrell and Eric Venbrux, Routledge, 2018, pp. 17–46.

Jedan, Christoph, Avril Maddrell, and Eric Venbrux. "Introduction: From Deathscapes to Consolationscapes: Spaces, Practices and Experiences of Consolation." *Consolationscapes in the Face of Loss: Grief and Consolation in Space and Time*, edited by Christoph Jedan, Avril Maddrell and Eric Venbrux, Routledge, 2018, pp. 1–13.

Kellert, Stephen R. "The Biological Basis for Human Values of Nature." *The Biophilia Hypothesis*, edited by Stephen R. Kellert and Edward O. Wilson, Island Press, 2013, pp. 42–69.

Kunkel, Adrianne Dennis, and Michael Robert Dennis. "Grief Consolation in Eulogy Rhetoric: An Integrative Framework." *Death Studies*, vol. 27, no. 1, 2003, pp. 1–38.

Lilley, Deborah. "New British Nature Writing." *Oxford Handbook Topics in Literature*, online ed., edited by Oxford Handbooks Editorial Board, Oxford Academic, 2013 [pdf pp. 1–18], https://doi.org/10.1093/oxfordhb/9780199935338.013.155. Accessed 2 Jan. 2024.

Meeker, Joseph W. *The Comedy of Survival: Studies in Literary Ecology*. Charles Scribner's Sons, 1974.

Morton, Timothy. *The Ecological Thought*. Harvard University Press, 2010.

Morton, Timothy. *Dark Ecology: For a Logic of Future Coexistence*. Columbia University Press, 2016.

Nash, Roderick Frazier. *Wilderness and the American Mind*. 1967. 5th ed., Yale University Press, 2014.

Neimanis, Astrida. "Hydrofeminism: Or, On Becoming a Body of Water." *Undutiful Daughters: New Directions in Feminist Thought and Practice*, edited by Henriette Gunkel, Chrysanthi Nigianni and Fanny Söderbäck, Palgrave Macmillan, 2012, pp. 85–100.

Oakley, Kate, Jonathan Ward, and Ian Christie. "Engaging the Imagination: 'New Nature Writing', Collective Politics and the Environmental Crisis." *Environmental Values*, vol. 27, 2018, pp. 687–705.

Oppermann, Serpil. "Rethinking Ecocriticism in an Ecological Postmodern Framework: Mangled Matter, Meaning, Agency." *Literature, Ecology, Ethics: Recent Trends in Ecocriticism*, edited by Timo Müller and Michael Sauter, Universitätsverlag Winter Heidelberg, 2012, pp. 35–50.

Pollard, Alexandra. "What the Growing Trend of Nature Memoirs Tells Us about the State of the World." *The Independent* (online), 26 January 2020. *ProQuest Central.*

Rigby, Kate. *Reclaiming Romanticism: Towards an Ecopoetics of Decolonization.* Bloomsbury Academic, 2021.

Scheffran, Jürgen, et al. "A Viable World in the Anthropocene: Living Together in the Common Home of Planet Earth." *Anthropocene Science*, vol. 3, 2024, pp. 131–142, https://doi.org/10.1007/s44177-024-00075-7. Accessed 5 Feb. 2025.

Slovic, Scott. "Environmental Humanities and the Public Intellectual." *Imaginative Ecologies: Inspiring Change through the Humanities*, edited by Diana Villanueva-Romero, Lorraine Kerslake and Carmen Flys-Junquera, Brill, 2022, pp. 15–30.

Spender, Stephen. *The Struggle of the Modern*. Hamish Hamilton, 1963.

Sullivan, Heather I. "The Dark Pastoral: Material Ecocriticism in the Anthropocene." *Ecocene: Cappadocia Journal of Environmental Humanities*, vol. 1, no. 2, December 2020, pp. 19–31, https://doi.org/10.46863/ecocene.16. Accessed 21 Feb. 2025.

The New Oxford Dictionary of English, edited by Judy Pearsall, Oxford University Press, 1998.

Thomashow, Mitchell. *Ecological Identity: Becoming a Reflective Environmentalist.* MIT Press, 1995.

Thomashow, Mitchell. *Bringing the Biosphere Home: Learning to Perceive Global Environmental Change.* MIT Press, 2001.

Tokarski, Mateusz. "Consolations of Environmental Philosophy." *Animals in Our Midst: The Challenges of Co-existing with Animals in the Anthropocene*, edited by Bernice Bovenkerk and Jozef Keulartz, Springer, 2021, pp. 445–467.

Tornøe, Kirsten Anne, et al. "The Challenge of Consolation: Nurses' Experiences with Spiritual and Existential Care for the Dying – A Phenomenological Hermeneutical Study." *BMC Nursing*, vol. 14, 2015, n. p., https://bmcnurs.biomedcentral.com/articles/10.1186/s12912-015-0114-6. Accessed 10 March 2024.

Totton, Nick, and Mary-Jayne Rust. Introduction. *Vital Signs: Psychological Responses to Ecological Crisis*, edited by Mary-Jayne Rust and Nick Totton, Routledge, Taylor & Francis Group, 2018, pp. XV–XXII.

Walton, Samantha. *Everybody Needs Beauty: In Search of the Nature Cure.* Bloomsbury Circus, 2021.

Wilson, Edward O. *Biophilia: The Human Bond with Other Species.* Harvard University Press, 1984.

Wilson, Edward O. "Biophilia and the Conservation Ethic." *The Biophilia Hypothesis*, edited by Stephen R. Kellert and Edward O. Wilson, Island Press, 2013, pp. 31–41.

Zalasiewicz, Jan, et al. "The Anthropocene: Comparing Its Meaning in Geology (Chronostratigraphy) with Conceptual Approaches Arising in Other Disciplines." *Earth's Future*, vol. 9, no. 3, 2021, https://doi.org/10.1029/2020EF001896. Accessed 4 Feb. 2025.

Zapf, Hubert. *Literature as Cultural Ecology: Sustainable Texts.* Bloomsbury Academic, 2017.

PART 1

(Re)Connecting with Nature as a Response to Modernity

∵

CHAPTER 1

Consolatio naturae in Joseph Conrad's *The Mirror of the Sea*

Sylwia Janina Wojciechowska

Abstract

Enumerated amongst the only "novelists in English worth reading" (Leavis 1), Joseph Conrad is perhaps best known for his psychological explorations of the human mind. This chapter, however, seeks to reassess Conrad's commitment to the world of real experience, and especially his concern with the broken connection with nature as experienced at the dawn of the twentieth century. Focusing on *The Mirror of the Sea* (1906), the chapter interprets Conrad's autobiographical narrative as a literary response to modernity and the transition registered by the discerning chronicler. It would seem that the critique of shrunken time/space relations and the breach of the unity between man and the sea allow *The Mirror of the Sea* to be regarded as an expression of disconcertment felt at the novel status of modern man, who, confronted with machines and scientific achievements, has apparently lost their sense of organic belonging to the natural world. A literary expression of what Stephen Spender calls "the modern necessity," Conrad's *The Mirror of the Sea* is claimed to employ "imagery which is of the time" to communicate the novelty of "the situation outside and beyond the [his] time" – a situation of a breach between man and nature. The adopted attitude of *prosochê*, that is that of serious contemplation, however, helps the writer express the centrality of the natural world in the modernised, and indeed perhaps even mechanised, reality. At times, *The Mirror of the Sea* ponders man as re-connected to nature in a set of images that represent the human being as ultimately belonging within the natural world rather than within machines and technology. This is viewed as the consoling potential of Conrad's maritime chronicle.

Keywords

Joseph Conrad – maritime chronicle – *prosochê* – attitudes to progress – organic connection between man and nature

© KONINKLIJKE BRILL BV, LEIDEN, 2026 | DOI:10.1163/9789004744110_003

Ever since Virginia Woolf's famous assertion that "human character changed" in or around 1910 (4), artists and scholars alike have attempted to pinpoint the onset of the change and to understand its nature.[1] However, neither Woolf nor academia were deluded as to the feasibility of the task as, according to Woolf, "the change was not sudden and definite [...]. But a change there was, nevertheless; and, since one must be arbitrary, let us date it about the year 1910" (4). The present paper proposes to investigate the significant change discerned by Woolf in terms of the relation of modern man to his/her natural surroundings, a relation then affected by both scientific developments and materialistic thinking. In the late-nineteenth-century literature, the man/nature bond had been presented as broken, or on the verge of breaking, by numerous writers, amongst them Thomas Hardy and Edward Thomas, and this line of argument was continued by modernists as well, such as D. H. Lawrence or T. S. Eliot. They all deplored a major shift in the priorities governing modern societies: apparently, technological advancement and material interests downplayed the significance of a life lived close to nature, which eventually left modern man disconcerted and frustrated. In his *Fantasia of the Unconscious* (1922), Lawrence exposes the broken connection with nature by stating: "we tend, through deliberate idealism or deliberate material purpose, to destroy the soul in its first nature of spontaneous, integral being, and to substitute the second nature, the automatic nature of the mechanical universe" (254). With their integrity and spontaneity dissolved, modern man internally transforms into a materialist, perhaps even an *automaton*, "and this is our danger to-day," warns Lawrence (254).

Like Lawrence and other writers of the day (see Spender 95–109; Seeber 267–275), Joseph Conrad also exposes the apparent predicament of humanity separated from its natural surroundings at the dawn of the twentieth century. Though generally sceptical about the human potential to improve on the march to modernisation, Conrad ventures to reassure his contemporaries with a series of consolatory images contained in his maritime autobiography, *The Mirror of the Sea*. Featuring Conrad's experiences at sea, the volume not only unmasks the processes of a gradual disintegration of the bond between man and nature, but also provides a certain consolation in the depiction of a seascape that features ships and sailors as natural components of the world rather than excluding them from his vision. The aim of the present chapter is to discuss the consoling potential of such a portrayal of the sea as a counterbalance to the images of accelerated modernisation, or the Lawrentian "mechanical

---

1   The title is a variation upon Boethius' *De consolatione philosophiae*, referenced further in the article.

universe," characteristically found in numerous narratives of the period. It is argued here that Conrad's commitment to nature in *The Mirror of the Sea* is twofold, not only featuring the sea in a series of maritime vignettes which depict its beauty, but also selecting this particular space as a point of reference for a discussion involving a serious concern regarding man's place in the world of nature. The images of fragmentation and technological advances which continue to re-appear in the narrative do not preclude the consolatory value of the volume as, by way of contradiction, they also draw attention to the importance of an attitude of attentiveness to the self, nature, and universe. Hence, the volume is barely escapist; rather, by retreating into the past in which man used to be closely connected to his natural surroundings, it emphasises the centrality of nature for modern man.

Published in 1906, *The Mirror of the Sea* fits well into the discussion on the nature of the change which occurred at the turn of the twentieth century and which affected the status of modern man. Not only is the publication characteristically "place[d] [...] about the year 1910" (Woolf 5), but the figure of its author, once a sailor, legitimises the significance of his claims regarding the natural world. As a former seaman, Conrad was perhaps the early modernist writer best placed to state his opinions on the transition recently effected within the man/nature bond as experienced at sea. His credibility was clear to his contemporaries (see Jabłkowska 189), who voiced their admiration for *The Mirror of the Sea* by praising the author's expertise with respect to nautical life. In a letter to Conrad, Henry James expresses his appreciation of Conrad's maritime narrative as follows:

> But the book itself [*The Mirror of the Sea*] is a wonder to me really – for it's so bringing home the prodigy of your past experience: bringing it home to me more personally & directly, I mean, the immense treasure & the inexhaustible adventure. No one has *known* – for intellectual use – the things you know, & you have, as the artist of the whole matter, an authority that no one has approached. (1 November 1906, qtd. in Stape and Knowles 58; italics original)[2]

The authority mentioned by James in the quotation above is that of a qualified sailor who knows how to translate his first-hand experience of the sea into literary terms. James recognises the duality of approach in *The Mirror of the Sea*,

---

2 Other appreciative readers included H. G. Wells, Ada and John Galsworthy, Edward Garnett, and Rudyard Kipling – see Stape and Knowles 52–59.

which, in Conrad's case, links the purely physical strain of sea praxis with the intellectual capacities of an artist.

Though published several years before Woolf's famous assertion regarding the transformation in the human character, the wind of the modernist change (Armstrong 47–53; Butler 1–24) flows in the narrative on several planes, beginning with the form of the narrative, loosely composed of vignettes and sea sketches, through its language and imagery, both animated by the principle of opposition, up to its generic hybridity. The autobiographical intent of *The Mirror of the Sea* is interspersed with the rich, fictional stratum of anecdotal stories and digressions enlivening the narration.[3] Though hardly uncommon in an autobiography (Depkat 280–286), the interlinking of fact and fiction in *The Mirror of the Sea* is remarkable in a narrative which was the novelist's first attempt at autobiographical non-fiction (Maunsell 18). In this context, however, far more engaging is the fundamental, conceptual duality, in which realistic, factual descriptions of the sea co-exist with paragraphs of fiction which, intermittently, border upon an escapist attitude in their re-imaginings of man as re-turned to the world of nature.

Promoting a life close to nature and in harmony with the natural surroundings, *The Mirror of the Sea* might have engaged with the pastoral. However, the narrative succeeds in commenting upon the pains of modernisation without deploying the pastoral disguise; rather, Conrad's engagement with the world of both nature and technology is straightforward and direct. I would argue that the directness is programmatic as it is announced in the motto to the volume which reads: "for this miracle or this wonder troubleth me right gretly [sic]" (Conrad, *Mirror of the Sea* 1). Taken from Boethius's philosophical treatise, *De consolatione philosophiae* (Book IV, Prose IV), the line apparently engages in dialogue with the title of the volume by playing with the polysemy of Latin *miraculum*: the Latin noun serves as the etymological source for both the Conradian "mirror" and "the miracle" as employed in the Boethian citation.[4] Whereas Zdzisław Najder focuses upon the Latin *miraculum* whose semantics he finds a pertinent reference to the title of the chronicle (95–96), I would also emphasise the significance of the entire quotation involving the verb, i.e.,

---

3 On the modernist autobiography, see DiBattista and Wittman 2014. On the fictional elements in Conrad's autobiography, see Said 1966; on its generic nuances, see Pacukiewicz 2012; Wojciechowska, *Nost/algia* 158–177. On the unusual frame of composition, see Skutnik 1978; on the particularities of the Conradian style, see Greaney 2015.

4 In classical Latin, the semantics of *miraculum* implies marvel and bewilderment generated by a thing that is not necessarily found admirable (Plezia 502, I and II).

"troubleth me." I agree with Najder that, in the narrative, the sea appears to be a mirror "[p]erhaps of Nature in general, of the universe" (95–96), but I would suggest that it primarily points to the "miraculous" change in the status of modern man as exposed to progress and industrialisation, which concerns the speaker "right gretly." Seen from this perspective, the natural world "mirrored" in the narrative is not the sole subject of scrutiny; instead, it becomes an object of reflection within a broader context of contemporaneity which embraces nature, man, and the world of technology. Such a broadening of the scope of reference for the Boethian motto is consistent with the philosophical under-pinning of Boethius's *De consolatione philosophiae* which generally promotes the practice of an attentive living known as *prosochê*, i.e., a particular concen-tration upon the self *within* the world (Ruszkiewicz 168–169).[5] As Ruszkiewicz observes, *prosochê* implies "the focus on the present moment [...] intended to encourage reflection on the place of the individuals within a wider scheme of things" (168). I would claim that in *The Mirror of the Sea*, the contemplative attitude of the autobiographical I embraces "a wider scheme of things" in so far as, while referring to technological progress, it promotes the attitude of an attentive living *within* the world of nature. As in the Boethian treatise, this con-templative stance, if accepted, may offer a certain *consolatio(em) naturae* to the readers by conveying an image of man linked to the cosmos, or, in Douglas E. Christie's words, by inviting them "to see and know the self *within* the world [...] as they are bound up together" (146).

Conrad's concern over the *miraculum* of progress separating man ever fur-ther from his/her natural surroundings is founded upon a persistent juxtaposi-tion of the past and the present: while it openly exposes the unprecedented haste and acceleration of life at the turn of the twentieth century (see Bauman 116–119; Wojciechowska, *Nost/algia* 76–91), it also includes a series of images which re-envision a world of which man once used to constitute a natural, organic part. The latter naturally creates the consoling potential of the volume: instead of being purely alarming, the conceptual dichotomy of the past-as-natural vs the present-as-mechanical inspires a contemplative interest in the readers who are invited to ponder the significance of nature for modern life. Consequently, the man/nature relation becomes the focus of serious reflec-tion which, subliminally, encourages the application of the Boethian *prosochê*, despite the fact that, at face value, the volume offers merely a range of compel-ling sea sketches. This, I argue, is a major achievement of *The Mirror of the Sea*

---

5 See also Douglas E. Christie, Chapter 1: "Immersion in the Larger Whole: Toward a Contem-plative Ecology," pp. 1–31, and Chapter 5: "*Prosoche*: The Art of Attention," pp. 141–178.

whose author, apparently, found a way to express what Stephen Spender calls "the modern necessity," that is the necessity to "express a situation outside and beyond the present time in imagery which is of the time" (98). Begun with the onset of the Industrial Revolution, long before Conrad's birth, and apparently continued until the present day, the march to modernisation as depicted in *The Mirror of the Sea* may be a particular instance of Spender's "situation outside and beyond the present time." I would argue that the range of images that feature sailing, on a ship or yacht, as an art embedded in the natural rhythm of life instantiates Spender's "imagery which is of the time." It is applied in *The Mirror of the Sea* in order to, first, inspire serious criticism regarding technological progress taking place not only at sea but ashore as well, and, second, to offer consolation in the visions of an organic connection between man and nature. While the passages commenting upon the automatic, or perhaps automatised, ways of modern living may alarm the readers, the fragments which envisage the natural rhythms of sailing counterbalance them, which, I argue, contributes to the consoling character of the entire narrative.

Symptomatically, the first sentence in the chronicle refers to "the rhythmical swings of a seaman's life" (Conrad 3), a life that is grounded in "your ship's routine, which I have seen soothe [...] the most turbulent of spirits. There is health in it, and peace, and satisfaction [...]; for each day of the ship's life seems to close a circle within the wide ring of the sea horizon" (Conrad 7).[6] Though the word "routine" may suggest a discouraging repetitiveness of actions, in the chronicle it is imbued with positive connotations linking human activities with the natural movements of the waves; thus, sailors are depicted as organically connected with the world of nature since their lives "borrow [...] a certain dignity of sameness from the majestic monotony of the sea" (Conrad 7). Thus, in the first few paragraphs in *The Mirror of the Sea* a close relation between man and nature is established – a healthy and organic connection which the human being is expected to create in their interactions with the external world. Characteristically, while arguing his case in Part 1, the author applies the semantics of health vs pathology, with the former consisting in "the soothing deep-water ship routine to establish its beneficent sway" (Conrad 7), and the latter cured with the healing monotony of sea life. Thus, the "imagery which is of the time" involves both the practice of sailing, and, by virtue of a metaphorical link between the sea and health, also the notion of therapy and convalescence. As I have argued elsewhere ("Pandemic(s)" 64), the therapeutic potential of

---

6 This matches the motto preceding Section 1, which reads: "And shippes by the brinke comen and gon,/ And in swich forme endure a day or two/ The Frankeleyn's Tale" (Conrad 3). On the Chaucerian practice of *prosochê* see Ruszkiewicz 170–176.

any narrative may involve the processes of self-identification; the consolatory intention in *The Mirror of the Sea* is no exception to the rule.

Against the somewhat soothing opening, Conrad's subsequent criticism regarding the processes of modernisation witnessed at sea at the turn of the twentieth century can be read as a dramatic contrast to the gentle images of life under sails. If the initial depiction of the routine regulated by the rhythms of nature and praised as "a great doctor for sore hearts and sore heads, too" (Conrad 7) sounds comforting, the images of advancing modernisation discernible in the maritime context create dissonance by featuring man's activities as contributing to the "regulated enterprise" of modern steamship (Conrad 30). After several pages depicting "the rhythmical swing[s]," the final statement closing sub-section VII, "The fine art [of sailing] is being lost" (Conrad 26), halts the reader midway in his/her admiration for sailing – the finality of the remark matches the abrupt brevity of the closing assertion.

Beginning with sub-section VIII, the speaker repeatedly alerts the reader to the sudden change in the man/nature relation that is perceptible at sea. By virtue of their dependence upon the elements, sailing ships used to enable a successful interaction between man and the world of nature which, however, is "already receding from us on its way to the overshadowed Valley of Oblivion" (Conrad 30). By contrast, the steamship has become a symbol of the substantial change in the status of man at sea. Founded upon such notions as punctuality, industry, and success (Conrad 30–31), the growing domination of the steamship suggests a rupture in modern man's relation to the natural surroundings of which he once was an organic part. The consistent deployment of the opposition between sails and steam seems to correspond to Spender's "imagery which is of the time" (Spender 98): the references to both types of ships endow Conrad's statements with a clarity and tangibility that can be understood by all, expert or layman. Since the speaker in *The Mirror of the Sea* recalls a number of sea voyages during which he had the opportunity to observe the seafaring practice of sailing ships and steamships alike, he sounds an expert when observing:

> your modern ship which is a steamship makes her passages on other principles than yielding to the weather and humouring the sea. She receives smashing blows, but she advances [...]. A modern fleet of ships does not so much make use of the sea as exploit a highway. The modern ship is not a sport of the waves. Let us say that each of her voyages is a triumphant progress; and yet it is a question whether it is not a more subtle and more human triumph to be the sport of the waves and yet survive, achieving your end. (Conrad 72)

In the quotation above, the breach between mankind and the natural world is conveyed on several levels. On the grammatical level, the possessive "your" openly re-affirms the chasm between "my/our" sailing expertise, as known in the past, which is juxtaposed with the dull activity of modern labour onboard a steamer. Likewise, the subordinate sentence, "which is a steamship," is a semantic pleonasm in a narrative whose major theme is the "steaming" modernisation juxtaposed with the natural ways of the sailing ship. Furthermore, the vocabulary in the passage connotes the world of technology, with such words as "advances," "exploit," "highway," and "a triumphant progress" that are strongly evocative of modern slogans. Closing the paragraph with a meaningful phrase, "achieving your end," the writer counters the idea of progress at all costs. Instead, he dismisses the modern connotations of achievement as financial gain (Conrad 137) and recalls that, in the times of sails, "achieving your end" used to mean the act of completing your sea voyage, a voyage performed in accordance with the natural rhythms of sea life.

As the narrative unfolds, the speaker's concern with impersonal technology and progress is communicated in ever more detail, with time and space as the major coordinates for the main sail/steam opposition around which the narration is constructed. Based on this juxtaposition – comprehensible even to laymen – the autobiographical *persona* draws conclusions exceeding personal reference: in fact, he forecasts a "situation outside and beyond the present time in imagery which is of the time" (Spender 98) when asserting that:

> the seamen of three hundred years hence will probably be neither touched nor moved to derision, affection, or admiration [for sailing]. They will glance at the photogravures of our nearly defunct sailing-ships with a cold, inquisitive, and indifferent eye. Our ships of yesterday will stand to their ships as no lineal ancestors, but as mere predecessors whose course will have been run and the race extinct. (Conrad 72–73)

The fragment quoted above reads as a serious warning against the processes of modernisation which distance the human being further and further from the world of nature, with the aftermath envisioned as progressing deterioration. The unnatural ways of modern man extend beyond the present moment, as the impassive reactions of the "cold, inquisitive, and indifferent eye[s]" of the future generations of seamen suggest. Allegedly, the steamship crew – "good men as they are, [who] cannot hope to know [the arcana of sailing]" (Conrad 72–73) – just efficiently manage the ship's machinery rather than traverse the waters by following the natural, pulsating rhythms of the elements. Such reasoning concurs with Spender's claims in *The Struggle of the Modern* (1963). Spender observes that "modern circumstances have set up a screen between nature and man so that the harmonious relationship realised in organic poetry in which the soul sees itself reflected in the physical environment, is prevented" (44). Given the musical and poetic quality of Conrad's *The Mirror of the Sea*,[7] Spender's observation seems relevant in the examination of Conrad's maritime chronicle. Like Spender, the speaker in *The Mirror of the Sea* openly regrets that "[t]he taking of a modern steamship about the world [...] has not the same quality of intimacy with nature" (Conrad 30); previously, sailing qualified as an art. The repeated exposure of the failings of the modern steamship strengthens the speaker's major claim about the vital importance of a life connected to nature for modern man.

Paradoxically, the constant juxtaposition of the natural and the mechanised coexists with the consoling aspect of the volume. *The Mirror of the Sea* emphasises the crucial significance of the world of nature – perhaps ever more significant in the context of modernity governed by science and technology. As the speaker navigates between the consoling and disconcerting images of seascape, the solace he offers appeals to the readers on, at least, two levels, namely through the sheer aesthetic beauty of the maritime vignettes, and also by presenting the space as endowed with certain qualities encouraging serious contemplation of the world. As has been indicated above, the opening of *The Mirror of the Sea* contributes to the former: not only does it depict the sea as a unique space which enables communion with nature, but it also evokes the feelings of awe and appreciation. Several poetic statements, such as "the setting of their sails resembles more than anything else the unfolding of a bird's wings" (Conrad 26), suggest a certain harmony between the sailor and the elemental forces that, momentarily, enables the reader to share the sailor's experience of an interaction with nature. Alongside other similar expressions which compare the sailing ship to "a sea-bird going to rest upon the angry waves" (Conrad 56), such images communicate the vitality of a strong bond between sailors and the natural forces. The narrative thus acquires a particular consoling and soothing quality which may persuade the laymen onshore to contemplate and re-assess the crucial significance of nature for human life.

Fenced by "the magic ring of the horizon" (Conrad 36), the sea denotes a space which is viewed as a whole – at times, as an atemporal whole. In such a context, "the days, weeks and months [imperceptibly elapse]" (Conrad 7). Freed from temporal ties and spatial constraints, the sea inspires reflection on human dependence upon the elemental forces, perceived with perspicuity during a storm:

---

7   Ada Galsworthy admires "its deep broad poetry" (a letter to Conrad, 1 October 1906, qtd. in Stape and Knowles 55) and Edward Garnett remarks on the "perfect phrases, every one of which is musical" (a letter to Conrad, 20 October 1906, qtd. in Stape and Knowles 56).

The greyness of the whole immense surface, the wind furrows upon the faces of the waves, the great masses of foam, tossed about and waving, like matted white locks, give to the sea in a gale an appearance of hoary age, lustreless, dull, without gleams, as though it had been created before light itself. (Conrad 71)

Unlike accounts romanticising the "adventurous voyages a[t] the Mediter-ranean" (Conrad 148), Conrad's description cited above features seascape as frightening and overwhelming. This agrees with Conrad's sound stance that the sea "has never been friendly to man" (Conrad 135). Apparently, its atem-porality welcomes the deliberate reflection announced in the motto in *The Mirror of the Sea* – "the wonder troubleth me right gretly [sic]" (Conrad 1). The disturbing passage quoted above invites contemplation since, "created before light itself," the sea appears as a unique space inspiring deep meditation about nature and civilisation. The speaker proceeds with the consideration of "the instinct of primitive man" (Conrad 71) and, finally, asserts that he speaks his mind as "the man [...] to whom the sea is not a navigable element, but an inti-mate companion" (Conrad 71). The emphasis placed upon the vital connection of human beings to the elemental forces stems from attentiveness and experi-ence rather than romantic fascination, and thus it involves both consolation and thoughtful contemplation upon the meaning of the world of nature for human life. On the whole, the entire Section XXII not only communicates the immediate dependence of man upon his/her natural surroundings, but sug-gests that first-hand experience of nature, both of its beauty and its extremity, instils a desire to ponder the fundamental importance of nature for the human being. In my view, Conrad argues his points by promoting the attitude of *prosochê*, i.e., attentive living, which, as Douglas E. Christie asserts, "is a way of seeing deeply into the *whole* of reality, God, the world, everything, and situat-ing oneself with integrity with relation to this whole" (142). It seems that Con-rad's volume negotiates the position of modern man *within* the world affected by progress. Instead of approving of the changes, the writer opens a discussion upon the significance of sustaining the connection to nature for modern man.

Such contemplative observations must have had a consolatory effect in times of accelerating modernity as, by virtue of the unhurried cadence of the narration, they counterbalanced the stress and haste of modernised life. Upon its publication in 1906, the reading public indeed appreciated both kinds of engagement with the external world in *The Mirror of the Sea*, namely the artis-tic rendition of the beauty of the sea and the deep concern with it as the arena upon which the sweeping changes induced by modernisation were traced. Besides Henry James, mentioned above, John Galsworthy also found the chronicle "magnificent [...] [and] rank[ing] with [...] [Conrad's] very highest work" (a letter to Conrad, 30 September 1906, qtd. in Stape and Knowles 54), asserting that it was "a splendid thing written from the heart of a true seaman, and full of truth and life and beauty" (a letter to William Archer, 29 September 1906, qtd. in Stape and Knowles 54). Similarly, Galsworthy's wife, Ada, admired "its manifold beauties" and "deep broad poetry [...] [which made her] feel as if [she] had been allowed to see soul-deep mysteries of the sea" (a letter to Conrad, 1 October 1906, qtd. in Stape and Knowles 55). The appreciation of the renowned literary couple and their admiration for Conrad's autobiographical narrative matches that felt by Rudyard Kipling, who openly expressed his gratitude for the volume, thus testifying to its consoling potential: "I read and re-read it all and thank you for it most heartily and gratefully" (a letter to Conrad, 9 October 1906, qtd. in Stape and Knowles 56). The same attitude is expressed in both Henry James's and William Rothenstein's letters, the latter stressing the transformative potential of the book. As its first readers assert, *The Mirror of the Sea* succeeded in featuring man as re-connected to the world of nature. Rothenstein thus described the process of an imaginary re-connection to nature which he allegedly experienced when reading the book: "[you] ha[ve] so enriched the world. [...] I walk about seeing masts & spars springing [...] into the air, & the waves of the sea are pursuing one another, blue & green & yellow & grey, embroidered with foam, across my vision" (a letter to Conrad, 1 November 1906, qtd. in Stape and Knowles 58).

In *The New Cambridge Companion to Joseph Conrad* (2015), Andrew Glazzard quotes a lengthy passage from a review for the *Daily News* which appeared after the publication. Its last few lines read:

> He [Conrad] is essentially creative in that he has learnt to apply the modern spirit of self-consciousness to the panorama of nature; he is doing in art what the transcendental idealists are doing in philosophy, teaching us to see material things as part of an all-embracing spiritual unity. (Scott James qtd. in Glazzard 59)

The "all-embracing spiritual unity" mentioned by Scott James connotes the practice of attention indicated in the Boethian motto to the volume – "for this miracle or this wonder troubleth me right gretly [sic]" (Conrad 1). Even though the opening sections of *The Mirror of the Sea* read as a series of gentle and romanticising sea images, which might suggest an escapist attitude of its financially troubled author, in fact, they never evade a serious concern with the novel status of modern man in the modernised world. This has been noted by Glazzard who detects "psychological and philosophical dimensions" to the static passages describing the beauty of seascape (59). I would continue on the same note by arguing that the maritime chronicle inspires not only an appreciation of its aesthetic value but, in the first place, of its attentiveness and serious commitment to nature. Instead of escapism, it communicates and promotes a deliberate living, marked by a serious consideration of the natural world. If escapism is one-directional, *The Mirror of the Sea* is a tale unfolding in several directions re-connecting the past with the present and man with his natural surroundings. Viewed from this angle, Conrad's narrative may grant a certain degree of consolation to those affected by the whirling storm(s) of progress, i.e., the ever more accelerating pace of life, the mechanised ways of contemporaneity, or the breach that has occurred between modern man and nature. In *The Mirror of The Sea*, nature is prioritised in order to outweigh the consequences of progress. Since, "[i]n his own time a man is always very modern" (Conrad 72), he/she must realise that his/her natural surroundings cannot be neglected. In my reading, this is the consolation granted by the *miraculum* featured in *The Mirror of the Sea:* it communicates an awareness of the substantial significance of the natural world for humanity.

Bibliography

Armstrong, Tim. *Modernism: A Cultural History.* Polity, 2005.

Bauman, Zygmunt. *Liquid Modernity.* Polity Press, 2017.

Boethius. *De consolatione philosophiae. The Consolation of Philosophy.* Translated by H. R. James, 2004. Downloaded from: *Project Gutenberg.* https://www.gutenberg.org/files/14328/14328-h/14328-h.htm. Accessed 9 Oct. 2023.

Butler, Christopher. *Early Modernism: Literature, Music, and Painting in Europe, 1900–1916.* Clarendon Press, 1994.

Christie, Douglas E. *The Blue Sapphire of the Mind.* Oxford University Press, 2012.

Conrad, Joseph. *The Mirror of The Sea: Memories and Impressions. A Personal Record: Some Reminiscences.* Dent and Sons, 1968.

Conrad, Joseph. *The Collected Letters of Joseph Conrad. Volume 3*, edited by Frederick R. Karl and Laurence Davies, Cambridge University Press, 1988.

Depkat, Volker. "Facts and Fiction." *Handbook of Autobiography/Autofiction*, edited by Martina Wagner-Egelhaaf, De Gruyter, 2019, pp. 280–286.

DiBattista, Maria, and Emily Wittman, editors. *Modernism and Autobiography.* Cambridge University Press, 2014.

Glazzard, Andre. "Letters and Autobiographical Writings." *The New Cambridge Companion to Joseph Conrad,* edited by J. H. Stape, Cambridge University Press, 2015, pp. 58–72.

Greaney, Michael. "Conrad's Style." *The New Cambridge Companion to Joseph Conrad,* edited by J. H. Stape, Cambridge University Press, 2015, pp. 102–115.

Jabłkowska, Róża. *Joseph Conrad: 1857–1924.* Zakład Narodowy im. Ossolińskich, 1961.

Lawrence, D. H. *Fantasia of the Unconscious.* Thomas Seltzer, 1922. Downloaded from: https://www.gutenberg.org/cache/epub/20654/pg20654-images.html. Accessed 4 April 2024.

Leavis, F. R. *The Great Tradition.* George W. Stewart, 1950.

Maunsell, Jerome Boyd. *Portraits from Life: Modernist Novelists and Autobiography.* Oxford University Press, 2018.

Najder, Zdzisław. *Conrad in Perspective: Essays on Art and Fidelity.* Cambridge University Press, 1997.

Pacukiewicz, Marek. "Cultural Aspects of Joseph Conrad's Autobiography. On the Digressive Structure of Some Reminiscences." *Yearbook of Conrad's Studies* (Poland), 2012, vol. 7, pp. 69–83.

Plezia, Marian. *Słownik łacińsko-polski. Tom III* [Latin-Polish Dictionary. Volume 3]. Wydawnictwo Naukowe PWN, 1998.

Ruszkiewicz, Dominika. "Prosochê and the Transformation of the Self in Geoffrey Chaucer's Poetry." *Perspektywy kultury,* vol. 45, no. 2, 2024, pp. 167–178.

Said, Edward W. *Joseph Conrad and the Fiction of Autobiography.* Harvard University Press, 1966.

Seeber, Hans Ulrich. *Idylle und Geschichte. Studien zur europäischen Idylle von Vergil bis W. H. Auden.* Universitätsverlag Winter, 2022.

Skutnik, Tadeusz. "O semantyce kompozycji *Zwierciadła morza.*" *O kompozycji tekstu Conradowskiego,* edited by Andrzej Zgorzelski, Wydawnictwo Morskie, 1978, pp. 7–28.

Spender, Stephen. *The Struggle of the Modern.* Hamish Hamilton, 1963.

Stape, J. H., and Owen Knowles, editors. *A Portrait in Letters: Correspondence to and about Conrad.* Rodopi, 1996.

Wojciechowska, Sylwia Janina. "Pandemic(s), Crisis, and Bibliotherapy." *Multidisciplinary Journal of School Education,* vol. 11, no. 1, 2022, pp. 51–69.

Wojciechowska, Sylwia Janina. *Nost/algia as a Mode of Reflection in the Autobiographical Narratives of Joseph Conrad and Henry James.* Peter Lang, 2023.

Woolf, Virginia. "Mr. Bennett and Mrs. Brown." *Hogarth Essays.* Hogarth Press, 1924, pp. 3–24. Downloaded from: https://archive.org/details/VirginiaWoolfMr.BennettAndMrs.Brown/page/n3/mode/2up. Accessed 22 Sept. 2023.

CHAPTER 2

Nature, Language and Silence in D. H. Lawrence's *St Mawr*

Nataša Tučev

Abstract

The chapter discusses D. H. Lawrence's novella *St Mawr* (1925) by referring to the ecocritical texts written by Christopher Manes (1992) and David Abram (1997). Both Manes and Abram stress the fact that in our culture nature has been silenced, whereas the status of a speaker has become an exclusively human prerogative. They contrast this state of affairs with the radically different attitude of animistic societies, for which non-human life forms were articulate and capable of communicating their presence and significance to humans. Abram considers the phenomenon of synaesthesia, arguing that it was of primary importance for animistic cultures, archaic man's self-knowledge and his embeddedness in the larger-than-human world of nature. The synaesthetic perception of the Other initiates the act of reciprocal participation, which may be regarded as an alternative language, one that is still available to the modern man. Lawrence's views expressed in *St Mawr*, and in a closely related non-fictional work "Pan in America" (1926), are very much in line with the ideas put forward by Abram and Manes. At the same time, they need to be interpreted in the context of general modernist preoccupations, the overwhelming sense of spiritual crisis in the early twentieth century Europe, and Lawrence's hope that a renewed interaction with nature may offer a possible solution. The chapter focuses on nature, language and silence as interrelated motifs in *St Mawr*, while also referring to the works of critics such as Keith Sagar (2005) and Jack Stewart (2011) to explore how Lawrence uses metaphor and literary synaesthesia to give voice to natural entities in the novel.

Keywords

D. H. Lawrence – ecocriticism – nature – language – speaking subject – animism – synaesthesia

1   Introduction

In his well-known ecocritical text "Nature and Silence" (1992), Christopher Manes points out that nature has been silenced and objectified in the dominant discourse of our culture, whereas the status of a speaking subject has become an exclusively human prerogative. Manes contrasts this attitude with the attitude of animistic societies, for which non-human life forms were inspirited and articulate, able to communicate their presence and significance to humans. In Manes' words, the archaic hunter-gatherers knew that the other, non-human languages must not be ignored: "In addition to human language, there is also the language of birds, the wind, earthworms, wolves, and waterfalls – a world of autonomous speakers whose intents [...] one ignores at one's peril" (Manes 15). Such entities in nature, however, are no longer regarded as autonomous speakers. Manes puts forward a theory on how having the status of a speaker is closely related to the moral considerations that we have for those who communicate with us; muting nature therefore directly leads to being able to exploit it with impunity (Manes 16).

Manes identifies the Middle Ages as the period when animism broke down in our culture, which he attributes, among other things, to the introduction of literacy and to Christian exegesis. Alphabetic writing enabled man to examine his own discourse in an abstract and generalized way, leading to the inference that meaning resided in human speech and not in the phenomenal world. Christian exegesis, on the other hand, was based on the principle that one should look beyond the literal meaning of Biblical passages, or *littera*, searching instead for their deeper significance indicating divine purpose and intention. This way of thinking was carried over into the realm of natural phenomena, which were likewise understood as mere *littera* – serving only to uncover the underlying spiritual revelations, instead of being appreciated or considered for their own merits. God was viewed as a transcendental subject speaking through birds, plants or animals. While the religious worldview is no longer predominant in Western culture, Manes highlights that from the Middle Ages onwards we have continued, in one form or another, to search for transcendental meanings when observing natural entities, still denying them autonomous voices (19–20). This is why he argues that, as a part of the ecological effort, it is necessary to find a language appropriate to environmental ethics: "we need to find new ways to talk about human freedom, worth, and purpose, without eclipsing, deprecating, and objectifying the nonhuman world" (Manes 24).

In a similar vein, David Abram (1997) stresses the invention of phonetic alphabet as one of the factors which significantly contributed to man's estrangement from nature. While the earlier developments of written signs, such as pictographs, still referred the reader to a field of meanings beyond the strictly human – for instance, by representing an animal, or alluding to one of its characteristics, such as speed or strength – the introduction of phonetic writing led to a more complete divorce from the natural world. The phonetic script merely transcribes the sounds produced by the human speech organs, so that the larger, non-human world is no longer a part of the semiotic (Abram 67).

Searching for an alternative way of communicating with nature which may still be available to modern man, Abram considers the phenomenon of synaesthesia and the concomitant notion of a chiasmic encounter with the Other. Synaesthesia is an experience we are familiar with on a daily basis, according to Abram, as we habitually "hear" the written words at the same moment when we see the letters on paper or screen. He maintains that the hunter-gatherers, in much the same manner, accomplished the convergence of various senses at the moment when they endeavoured to "read" the phenomena they encountered in nature (Abram 79). In fact, as he argues, synaesthesia is still central in our perception of others. Our experience of the world is largely synaesthetic: the visible appearance of a cat, or a tree, for instance, does not register in our mind separately from the sounds they produce, nor from their texture, or the olfactory impression they make on our senses. The entity we perceive, as Abram points out,

> is precisely the place where these separate sensory modalities join and dissolve into one another [...] The differentiation of my senses, as well as their spontaneous convergence in the world at large, ensures that I am a being destined for relationship: it is primarily through my engagement with what is *not* me that I effect the integration of my senses, and thereby experience my own unity and coherence. (Abram 80)

In this way, as Abram argues, perceiving the Other is essentially connected with self-knowledge: members of early oral communities, having no formal writing systems, came to know themselves and gain a sense of coherence and unity primarily by engaging with animals and the animated landscape. These entities offered one a kind of mirror, or a reflection of oneself, by bringing together all sensual impressions in the acts of interaction and perception (Abram 79).[1]

---

1   Abram's explanation of this process bears some similarity to the well-known Lacanian concept of the "mirror stage," whereby in early infancy a child gains a sense of a unified self by observing its image in a mirror. According to Lacan, the mirror situation also extends to other

Abram also discusses a closely related term of *chiasm* which originally referred to the brain region where the neuronal fibres from the left and right eye cross and interweave, resulting in a single vision. However, he argues that it is also possible to talk about a chiasm "between the various sense modalities," as well as a chiasm in the sense of some kind of middle ground where we encounter other beings in nature in the act of reciprocal participation and dynamic exchange (81). This exchange which we establish with the Other, by focusing on it with our various senses and experiencing their convergence, may be regarded as an alternative language which man can still use to communicate with nature.

The discussion in D. H. Lawrence's essay "Pan in America," written in 1926, is very much in line with contemporary ecological concerns, and a number of parallels may be drawn with the ideas put forward by Abram and Manes. Similarly to these two authors, Lawrence points to the importance of animistic cultures and their awareness that the non-human entities have their own utterances and discourse. He presents it in the essay as the awareness of Pan, a Greek deity which for him symbolizes vital energy and the mystery of being in all living things. However, Lawrence links Pan not only to the ancient Greek myths, but also to the beliefs of Native Americans, which he describes as essentially animistic (Ehlert 114). A Native American hunter, as Lawrence maintains, "suffered and hungered," but at least lived "in a ceaseless living relation to his surrounding universe" ("Pan in America" 163), deeply alert to the presence of all living forms in his environment and their influence; he established a kind of mutually beneficial communication which enabled him to survive and thrive among them.

According to Lawrence, what marked the end of such communication in our culture was man's increasing desire for dominance over the phenomenal world: historically, as he argues, the animistic attitude to nature gave way to man's focus on inventing "engines and instruments which should intervene between him and the living universe, and give him mastery" ("Pan in America" 162). Lawrence discerned the end results of this harmful human tendency towards mastery in his own age. In "Pan in America," he comments on how Western man at the beginning of the twentieth century "insulates himself more and more into mechanism," which in turn damages not only nature, but human psyche as well, leading to a state of inward inertia and loss of vitality (160). While Lawrence does not advocate some idyllic return to the pre-industrial era, he proposes that this controlling attitude to nature needs to be

---

objects and persons – in which, at this early stage of mental development, we may likewise find our integrated self-image reflected back to us (Eagleton 164–165).

abandoned, and our doors of receptivity re-opened. This may result, as it often does in Lawrence's fiction,[2] in hopeful human–nature encounters, transcending the egocentric and human-centred approach to animals and other natural entities, and in healing the fragmented, reified state of modern man's spirit.

While these ideas are consistent with the ecocritical concerns of Abram and Manes, an important difference lies in the way Lawrence approaches the theme of silence in conjunction with nature. Thus, Manes stresses the fact that nature has been silenced in our culture, and that this has led to its being objectified and exploited; on the other hand, Lawrence is more focused on the inadequacies of human language. His statement that "speech is the death of Pan" ("Pan in America" 160) implies a sense of mystery inherent in nature which surpasses conceptual, rational thinking, and therefore cannot be expressed by means of language; "the Pan silence," as we are told in the essay, is "full of unutterable things" (161).

Lawrence takes the same approach in his novel *St Mawr* (1925). As the following analysis will aim to demonstrate, the novel thematizes human language and its inability to communicate with the non-human subjects in nature, but also engages in a quest for alternative forms of communication. In his study *D. H. Lawrence: Language and Being* (1992), Michael Bell argues that these two levels of preoccupation with language always co-exist and interact in Lawrence's writing: his characters repeatedly reflect on language, but so does the author himself, which is evidenced in his approach to narration and style (Bell 4). For this reason, the following sections of the essay will explore nature, silence and language as interrelated motifs in *St Mawr*, but also discuss how Lawrence uses metaphor and literary synaesthesia to give voice to natural entities in the novel. These stylistic and thematic concerns should also be understood as a part of Lawrence's endeavour to restore relations with the natural world and offer a cure for the alienated human condition.

2   The Aura of Silence and the Speech of Nature

The plot of *St Mawr* focuses on Lou Carrington, a young American expatriate living in London, and her spiritual quest in which interacting with nature plays a crucial part. Lou's first encounter with the stallion St Mawr affects her profoundly; it initiates the desire for a more comprehensive knowledge of

---

2   In addition to *St Mawr*, some notable examples of restorative and invigorating encounters with nature may be found in Lawrence's novels such as *Sons and Lovers* (1913), *The Rainbow* (1915) and *Lady Chatterley's Lover* (1928).

the self, but also causes her to feel estranged from the superficial life of the metropolitan elite to which she has hitherto belonged. Her marriage to Rico, a young Australian aristocrat and talentless painter, also begins to strike her as meaningless, lacking passion and sincerity. In a crucial incident, Rico is injured due to his own foolish and arrogant actions while riding St Mawr. His friends want to either kill or geld the stallion; Lou, however, realizes that St Mawr is not to blame and decides to save him by taking him to her native Texas. Once there, she feels an urge to withdraw even further from civilization. Eventually, she buys a ranch in the mountains of New Mexico, where she plans to live in isolation for a while, hoping to establish rapport with the spirit of wild nature, and also heal her inner being in the process.

The motif of communicating with nature is introduced in the pivotal early scene in the novel, when Lou first lays eyes on the magnificent stallion St Mawr. The stallion's vitality and energy, and a sense of mystery surrounding his entire bearing, make a profound impression on her. She has a feeling that she is in the presence of some pagan deity or a demon, who is asking her a question which she cannot quite articulate:

> The wild, brilliant, alert head of St Mawr seemed to look at her out of another world. It was as if she had had a vision, as if the walls of her own world had suddenly melted away, leaving her in a great darkness, in the midst of which the large, brilliant eyes of that horse looked at her with demonish question [...] What was it? Almost like a god looking at her terribly out of the everlasting dark, she had felt the eyes of that horse; great, glowing, fearsome eyes, arched with a question, and containing a white blade of light like a threat. What was his non-human question, and his uncanny threat? (*St Mawr* 30–31)

Lou's ability to establish an exchange with St Mawr and hear his "non-human question," even though she is not capable of translating it in conceptual terms, may be ascribed to what Michael Bell calls her "mythic sensibility." Namely, as Bell argues, the characters in Lawrence's novels display three basic modes of sensibility, which, drawing on Cassirer's philosophical anthropology, he terms mythic, religious and scientific. The mythic sensibility was characteristic for the archaic, animistic societies, its main feature being a sense of continuity which appeared to exist between the external world of nature and the inner world of human feelings. In this mode, which Bell also calls pre-dualistic, nature itself is experienced as divine (in the same manner that Lou experiences St Mawr as a kind of totemic divinity). The religious mode, as Bell explains, "is characterised by the dualising of natural and supernatural, as well as the duality of self and world" (79); in this phase, the divine is experienced as a quality transcending nature and standing in opposition to it. The third mode is secular, viewing nature with analytic objectivity, "to be possessed as an idea separate from man" (Bell 3). Bell also indicates that each of these modes is embodied in a different kind of language in Lawrence's writing. The main difference between Cassirer's categorization and the one we encounter in Lawrence's novels, as he explains, is that for Cassirer these modes are evolutionary, and pertain to certain historical phases in the development of human consciousness, whereas for Lawrence they can exist simultaneously in a culture, or even within a single individual: "In Lawrence's conception," as he says, "the phases of the past persist potently" (Bell 84).

Bell's insights may also be compared to those in Manes' essay, regarding the treatment of nature through history: in animistic societies, nature was inspirited and had the status of a speaking subject; in the Middle Ages, God was viewed as a transcendental speaking subject, merely using natural phenomena as signs that need to be deciphered; whereas the modern era, for both Manes and Lawrence (as interpreted by Bell), is secular but still characterized by our tendency to search for transcendental meanings, denying non-human beings their subjectivity and autonomous voices (Manes 19–20). One possible interpretation, therefore, of the scene between Lou and St Mawr is that at the moment of the encounter powerful emotions cause Lou to experience a different mode of consciousness, reminiscent of mythic or animistic sensibility, which enables her to communicate with a natural entity.

It is also important to notice that in the above quoted passage Lawrence resorts to the use of free indirect style of narration. As Randall Stevenson explains, this often happens in Lawrence's writing when his characters are shown dealing with turbulent emotions and liminal psychic states which they cannot fully verbalize, nor express in rational terms. In such cases, transcribing the "inner voices" of characters – by using the stream-of-consciousness technique, for instance – would not represent them adequately, because the reader would not gain access to the portion of their experience which remains beyond articulation (Stevenson 20). In this context, Lou's striking encounter with St Mawr may be viewed as a kind of dialogue with nature which takes place outwith rational discourse, and therefore cannot be reported directly.

Lawrence also resorts to metaphorical language in order to explain Lou's experience, stating that "the walls of her own world had suddenly melted away" when she first saw the stallion (22). The wall may refer to a barrier separating our ego-consciousness from the deeper psychic layers, or separating the rational self from the vital instinctive energies in our being, to which Lawrence ascribes utmost importance. In his non-fictional work, *Psychoanalysis and the*

Unconscious (1921), he postulates a theory of the existence of a "non-Freudian," or "pristine" unconscious, which he explains as the life instinct itself, residing in all living beings and ontologically preceding consciousness. According to Lawrence, the only way to live spontaneously and creatively is to get in touch with this deep core of the self: "We must discover, if we can, the true unconscious, where our life bubbles up in us, prior to any mentality. [...] It is the spontaneous origin from which it behoves us to live" (*Psychoanalysis and the Unconscious* 15). This vitalist position also determines Lawrence's criticism of his contemporary society, whose members, in his view, tend to identify only with their rational ego and therefore live sterile lives devoid of energy and passion. In *St Mawr*, Lawrence explores whether the encounters with nature and natural entities (such as Lou's encounter with the stallion) can result in the melting of psychic barriers, making it possible for his contemporaries to reconnect with the natural energies in their own being and experience life "straight from the source," as he terms it in the novel (*St Mawr* 61).

This also explains why Lou starts to feel increasingly estranged from her husband Rico and from the London upper-class society in general. While attending dinners, cocktails and dances, she has the impression that she is surrounded by "young bare-faced unrealities" (42), wraiths who are only ostensibly alive. As the narrative suggests, the reason these people have such meaningless, two-dimensional lives is that they are completely divorced from the deeper layers of their psyche – the very layers that Lou became aware of in the presence of St Mawr (Tučev 53). For the same reason, Lou thinks of Rico as "bodiless" and as identifying only with his "head," i.e., his conscious, rational self. Rico's head, clean-shaven and "perfectly prepared for social purposes," is described in the novel as one of the "talking heads of modern youth" (34). As the phrase suggests, Lawrence wishes to indicate that these young people's lack of vitality also affects their use of language. Their minds seem to wither along with their bodies, resulting in a lack of spontaneity and creativity in thinking and speaking. In an exchange with her mother, Mrs Witt, Lou likewise complains that the conversations of the metropolitan elite are repetitious and "deadly dull," like "knitting the same pattern over and over again" (60). Another example of the failure of language is a scene in which Mrs Witt watches a funeral in Shropshire; it causes her to realize that her entire life may be summed up in a few shallow phrases about her social persona exchanged by the members of the high society. Her fear is that all that will be left of her is "a daily sequence of newspaper remarks" (93) and gossip of the upper-class circles. Generally, human language in Lawrence's novel is shown as suffering the consequences of having severed its vital link to the discourse of the natural world.

In contrast to the human conversations, which are presented in the novel as dull and repetitive, drawing on established rational concepts and conventional attitudes, the silence of nature appears to convey a much deeper significance. The inaudible interaction that takes place between Mrs Witt's servant Lewis and the animals he tends to strikes Mrs Witt as a mystical exchange pregnant with meaning beyond her reach. As she watches him tend to an injured bird, or groom St Mawr, she becomes aware of

> another communion, silent, excluding her. [...] another world, silent, where each creature is alone in its own aura of silence, the mystery of power [...] The visible world, and the invisible. Or rather, the audible and the inaudible. She had lived so long, and so completely, in the visible, audible world. She would not easily admit that other, inaudible. (104)

In the same way as Lou, Lewis seems to be possessed of mythic sensibility which enables him to converse with non-human entities. This is also evident in the scene in which he conveys to Mrs Witt his belief that he can hear the farewell of the sinking moon or the speech of the trees as he passes them by. Commenting on these motifs, Marina Ragachewskaya points out that Lawrence makes Lewis's character "inhabit a different world from that of the rest of humans" (par. 15). Mrs Witt, on the other hand – despite gradually becoming aware of the inadequacies of language, and the existence of an entire realm of being beyond its reach – remains confined to the analytical mode of consciousness which is incompatible with Lewis's animism. As Michael Bell points out, when Lawrence's characters display incompatible modes of sensibility, this often represents an insurmountable obstacle to their getting together, even if there exists mutual attraction (Bell 84). This may be the reason why Lewis turns down Mrs Witt's unconventional marriage proposal.

Another significant example of dialoguing with nature occurs in the part of the novel where Lou moves to a secluded New Mexico ranch. She is shown to communicate with the "wild spirit" which resides in the place in much the same manner as she was able to communicate with St Mawr at the beginning of the novel. In another related essay, titled "The Spirit of Place,"[3] Lawrence expresses a belief that the locality itself exerts influence on its human inhabitants, marking them in a unique way. Some of the terms he uses in the essay to explain the agency of this spirit, such as "effluence," "vibration," "chemical exhalation" or "terrestrial magnetism" ("Spirit of Place" 17) may be regarded as

---

3   "The Spirit of Place" was published as the introductory chapter to Lawrence's *Studies in Classical American Literature* (1923).

a kind of language; they suggest, in terms used by Manes, that Lawrence views the terrain as a speaking subject, relaying its presence and significance to the dwellers. In *St Mawr*, Lou likewise experiences the New Mexico landscape as vibrantly alive and inspirited, addressing her: "It's a spirit. And it's here, on this ranch. It's here, in this landscape. It's something more real to me than men are, and it soothes me, and it holds me up. [...] It's something wild, that will hurt me sometimes and will wear me down sometimes" (155). As Lou's words imply, the "spirit" of this particular place is harsh and at times even hostile to those who wish to settle there; nevertheless, the wild New Mexico scenery is presented in the novel as breathtakingly beautiful, in descriptive passages which also serve to provide a contrast to the meanness and pettiness of human affairs and the lives of the wealthy English elite presented in the previous sections. Despite the harshness of wild nature, Lou views the ranch as a place of restoration and healing where she will recover from the experience of the exhausted European civilization, which she now considers damaging to her soul (Tučev 60).

3   Interspecies Metaphors and Literary Synaesthesia

As Bell points out, "There is almost invariably in Lawrence's novels a signifi-cant interaction between the narrative language of the given book and the way language is thematised in it" (4). The same is true of *St Mawr*: in addition to thematizing language and its complex relationship to nature in the novel, Law-rence also uses the narrative language itself in an effort to restore the sense of nature as a living and speaking subject. In his study *Literature and the Crime Against Nature* (2005), Keith Sagar claims that Lawrence accomplishes this pri-marily through his use of metaphors. Arguing that creative imagination has been instrumental in keeping the animistic consciousness alive "through the thousands of years of its general rejection and persecution" (370), Sagar also adds that the language it employs for that purpose is essentially metaphorical: "Metaphor is the linguistic equivalent of touch. It is the link, the bridge, the meeting, the marriage, the atonement, bit by bit reconstructing the world as a unity, blissfully skipping over the supposed chasms of dualism" (371). Jack Stewart singles out Lawrence's interspecies metaphors in particular, saying that they "surprise and delight as they bridge the gap between human life and birds, beasts and flowers" and reflect the author's quest for eco-monism (93).[4]

---

4   Eco-monism, in the words of Donald Gutierrez, "can be defined as an ecological sense of human unity with nature and the earth, one in which the recognition of a crucial interde-pendence between humanity and nature could help preserve nature and thus humanity"

In *St Mawr*, Lewis in particular is portrayed as one such character who bridges the gap between the human and the natural world, which Lawrence implies through the use of metaphoric language (Humma 50). Mrs Witt, for instance, reflects on the way Lewis rides St Mawr: "He seems to sink himself in the horse. When I speak to him, I'm not sure whether I'm speaking to a man or to a horse" (38). Interspecies metaphors are also used to convey Lewis's physical appearance: in opposition to Rico's clean-shaven face, Lewis is depicted as staring "from between his bush of hair and beard [...] like an animal from the underbush" (*St Mawr* 34) and his eyes are likened to those of a wild cat peering from darkness (Humma 51). The metaphors Lou uses to describe the kind of man who would be her ideal partner – a mystical man, endowed with intuitive knowledge and a deep sense of connectedness with the non-human world – are likewise interspecies metaphors. He would be "a pure animal man," "as lovely as a deer or a leopard, burning like a flame straight from underneath. And he'd be part of the unseen, like a mouse is, even. And he'd never cease to wonder, he'd breathe silence and unseen wonder, as the partridges do" (62). In the given passage, John Humma also draws attention to the use of words such as "underneath," "unseen" or "silence" which Lawrence associates with natural entities throughout the novel. Humma argues that Lawrence establishes a dichotomy between the human and the animal through pairs of opposite signifiers, such as surface-subterranean, outer-inner, higher-lower, visible-invisible, light-dark, heard-unheard (50). However, as seen in the above passage, in certain contexts Lawrence describes human characters by employing attributes pertaining to animals, and in this way suggests the possibility of transcending the dichotomy, or bridging the gap between the two.

Another stylistic method by which Lawrence invokes the sense of nature as a speaking subject in *St Mawr* is literary synaesthesia. As discussed above, the sensory experience of synaesthesia was of primary importance for the animistic cultures, and is still available to the modern man. David Abram argues that synaesthetic perception of the Other initiates the act of reciprocal participation, which may be regarded as an alternative kind of language, the one that could still enable us to communicate with the non-human world (81).

In "Pan in America," Lawrence documents a striking example of this kind of communication. As he explains in the essay, during his stay at a ranch in New Mexico he became keenly aware of the surrounding landscape, and the imposing presence of a large pine tree in front of his cabin. The following passage

---

(39). It may also be related to Bell's contention that some of Lawrence's characters exhibit a pre-dualistic mode of consciousness whereby the internal world of human feelings and the external world of natural phenomena are perceived as inseparable (Bell 79).

deals both with the author's synaesthetic perception of the tree and with his impression that a mystical encounter took place whereby the man and the tree penetrated each other's lives:

> It vibrates its presence into my soul, and I am with Pan. Something fierce and bristling is communicated. The piny sweetness is rousing and defiant, like turpentine, the noise of the needles is keen with aeons of sharpness. In the volleys of wind from the western desert, the tree hisses and resists. [...] I have become conscious of the tree, and of its interpenetration into my life. [...] I am even conscious that shivers of energy cross my living plasm, from the tree. [...] And the tree gets a certain shade and alertness of my life, within itself. ("Pan in America" 158)

In his analysis, Jack Stewart also points out that Lawrence's perception in the passage is synaesthetic, participatory and essentially animistic: his response to the tree "fuses sight, smell, touch, and sound in a totemic complex, foregrounding ontological qualities of presence, oneness, fierceness, defiance, sharpness, resistance, and self-assertion" (96).

Lawrence uses the same imagery in *St Mawr*, towards the end of the novel, when he depicts a secluded ranch to which Lou has decided to withdraw in order to establish contact with wild nature.[5] The central position in the landscape – and, by implication, in Lou's experience as well – is assigned to a huge pine tree growing close to where she lives: Lou views it as the guardian of the place. Lawrence fuses disparate sensory data in a synaesthetic description of the tree in order to represent the powerful impact it has on Lou's emotions and imagination. The tree is a "great pillar of pale, flakey-ribbed copper"; sight and touch combine to invoke in Lou a feeling that it conveys "strange, callous indifference" and "grim permanence." These are coupled with olfactory impressions of the tree's "cold, resinous sap surging and oozing gum" and the auditory perception of the "wind hissing in the needles" (144). Diverse sensory modalities join and dissolve into one another, resulting in the experience of reciprocal participation. At the narrative level, Lawrence's synaesthetic description of the tree suggests to the reader that Lou experiences it as a speaking subject and engages with it in an act of wordless communication. At the same time, the wild New Mexico landscape serves as a connecting factor, a place where the chiasmic intertwining of the two life-forms can occur (Stewart 96).

---

5   The description of the Las Chivas ranch in *St Mawr* is to a great degree based on the biographical facts and coincides with the description of the Kiowa Ranch in New Mexico, where Lawrence and his wife lived in 1923 and 1924 (Ehlert 115).

In Lawrence's novel, this chiasmic encounter leads not only to Lou's renewed ability to understand the speech of nature, but also to her experience of comfort and psychological recovery.

4   Conclusion

Even though it was written and published a century ago, Lawrence's novella *St Mawr* foreshadows many of the contemporary concerns regarding the natural world and a quest for an alternative language based on environmental ethics, which would recognize nonhuman entities as autonomous speaking subjects. Using the metaphor of Pan, a Greek deity which for him signifies "the hidden mystery – the hidden cause" (*St Mawr* 65) in all living things, Lawrence argues that human speech, having severed its link to nature, marks the death of Pan. This entails that man no longer engages in a vital exchange with animals and the animated landscape; however, it also suggests Lawrence's recognition of a sense of mystery in nature which cannot be expressed by means of language. *St Mawr* thematizes the inadequacies of language – through motifs such as Lou's disappointment in the "talking heads of modern youth" (34) and their repetitive conversations, or Mrs Witt's fear that her life will be reduced to a sequence of newspaper remarks – while, on the other hand, it represents the "Pan silence" of nature as pregnant with meanings beyond human reach. These meanings may still be revealed to some individual characters in the novel, such as Lou or Lewis, whose mythic sensibility enables them to approach the stallion St Mawr, or a pine tree in the New Mexico wilderness, with the full awareness of their vibrant lives and autonomous agency. In this context, it is important to point to Michael Bell's remark that for Lawrence animism is not simply a stage in the evolution of human consciousness confined to the past, but a mode of thinking and feeling that is still available to modern man, making it possible for some of the characters in *St Mawr* to engage in a meaningful dialogue with nature.

Examining Lawrence's own language also leads to important insights, as he endeavours to re-create animistic awareness at the level of narration and style. Critics such as Jack Stewart (2011), Keith Sagar (2005) or John B. Humma (1990) have highlighted the importance of Lawrence's interspecies metaphors, which invoke the archaic, pre-dualistic attitude to nature and nonhuman life forms. Lewis, the character in *St Mawr* who most strikingly bridges the gap between the human and the animal, is depicted through the use of such metaphors. Lou, on the other hand, reaches for metaphorical language herself in her ruminations about the mystical man who would partake of the qualities of various beings in nature. Lawrence's use of free indirect style throughout the novel also suggests the otherness of the speech of nature: his characters, such as Lou or Mrs Witt, appear to hear it at the very threshold of consciousness and are not capable of fully verbalizing it, which is why Lawrence resorts to merging the voice of the narrator and that of a character in order to represent such experiences. As for the use of literary synaesthesia in the novel, it echoes Lawrence's own psychological experience documented in "Pan in America," where the convergence of various sense modalities enabled him to have a totemic vision of a pine tree as a "guardian" of the New Mexico landscape. This biographical material is transposed almost seamlessly to the novel, where Lou engages with a pine tree in the same manner, her synaesthetic experience leading to a kind of wordless dialogue with the natural world and a new insight about the self. In this sense, Lou repeats the experience of archaic oral cultures, whose members gained self-knowledge through an act of reciprocal participation and engagement with their environment; thus, through his protagonist, Lawrence conveys his belief that this mode of consciousness is still accessible in his own age.

Given Lawrence's view that the modern man "insulates himself more and more into mechanism" ("Pan in America" 160), which implies both the objectification of nature and the alienation of the human individual, his novel offers a hopeful vision on how both conditions can be remedied. Dialoguing with nature is thus represented by him not only as a way of establishing environmental ethics, but also as a possible cure for the Western man's spiritual crisis, one deeply felt at the beginning of the twentieth century.

Bibliography

Abram, David. *The Spell of the Sensuous: Perception of Language in a More-Than-Human World*. Vintage Books, 1997.

Bell, Michael. *D. H. Lawrence: Language and Being*. Cambridge University Press, 1992.

Eagleton, Terry. *Literary Theory: An Introduction*. University of Minnesota Press, 1989.

Ehlert, Anne O. *There's a Bad Time Coming: Ecological Vision in the Fiction of D. H. Lawrence*. Uppsala University Library, 2001.

Gutierrez, Donald. "D. H. Lawrence's 'Spirit of Place' as Eco-monism." *D. H. Lawrence: The Journal of the D. H. Lawrence Society*, 1991, pp. 39–51.

Humma, John B. *Metaphor and Meaning in D. H. Lawrence's Later Novels*. University of Missouri Press, 1990.

Lawrence, D. H. "Pan in America." *Mornings in Mexico and Other Essays*. Cambridge University Press, 2009, pp. 153–164.

Lawrence, D. H. *Psychoanalysis and the Unconscious and Fantasia of the Unconscious.* Cambridge University Press, 2004.

Lawrence, D. H. *St Mawr and Other Stories.* Cambridge University Press, 1987.

Lawrence, D. H. "The Spirit of Place." *Studies in Classical American Literature.* Cambridge University Press, 2003, pp. 13–19.

Manes, Christopher. "Nature and Silence." *The Ecocriticism Reader: Landmarks in Literary Ecology*, edited by Cheryll Glotfelty and Harold Fromm, University of Georgia Press, 1996, pp. 15–29.

Ragachewskaya, Marina S. "Horses, Women, Storms and More: The Dialogics of the Human and Non-Human in *St. Mawr.*" *Études Lawrenciennes*, vol. 53, 2021. https://doi.org/10.4000/lawrence.2830.

Sagar, Keith. *Literature and the Crime Against Nature.* Chaucer Press, 2005.

Stevenson, Randall. *Modernist Fiction: An Introduction.* Harvester Wheatsheaf, 1992.

Stewart, Jack. "Flowers and Flesh: Color, Place, and Animism in "St Mawr" and "Flowery Tuscany." *The D. H. Lawrence Review*, vol. 36, no. 1 (Spring–Summer), 2011, pp. 92–113.

Tučev, Nataša. *An Introduction to the Modernist Novel.* Faculty of Philosophy, University of Niš, 2021.

CHAPTER 3

Crossing the Rift: Interwar Intellectuals and Mass Escapism to Nature

Ladislav Vít

Abstract

The discovery of rural areas by the masses in interwar Britain significantly influenced the cultural atmosphere of the time. It also compelled the intelligentsia to ponder the open-air ethos and reconsider the connection between humans and the natural world. This chapter shows that while some intellectuals embraced the idea of nature as a consoling place, others denounced it sharply as a sign of romanticized objectification and naïve myth enticing the public into escapism and away from their civic responsibility. First, the cultural aspects of rural activities are delineated with a focus on the development of an approach to nature as a means for fostering citizenship, improving morals, and facilitating mental and physical recuperation. The chapter then examines the work of three authors who shared criticism of the outdoor movement and open-air ethos while differing in their opinion on the ability to attain consolation through a state of communion with nature. It attends to the specificities, nuances and reservations characterizing Henry Scott Holland's and George Macaulay Trevelyan's belief in the possibility of bridging the rift between man and nature for the sake of attaining bodily recreation and spiritual consolation. Their position is then contrasted with a detailed analysis of the dissenting views held by Wystan Hugh Auden, one of the leading intellectual voices of the period with a remarkable yet largely unexplored spatial sensibility and strong views on the relation between humans and the natural world. Because of his stature, Auden's unwillingness to view nature as a consolatory hideaway represents a vital counterweight to voices promoting the possibility of a recuperative communion with the natural world. Also, aware of the worsening international situation spiralling towards war, Auden provides a clear critique of the open-air movement in his 1930s writing and debunks it as a socially irresponsible phenomenon caused by spatial mythologizing. As such, this chapter shows how the vogue for rural pursuits generated an impressive variety of responses and so became a crucial component of interwar politics and the cultural atmosphere.

Keywords

nature – escapism – interwar Britain – H. S. Holland – G. M. Trevelyan – W. H. Auden

．．
．

When on April 8, 1939, the *Daily Worker* exhorted: "Let's All Go to the Country –
It's Ours!," the fascination of the masses with rural sojourns was in full swing.
Three years earlier, Clive Rouse had already questioned the thematic relevance
of his *The Old Towns of England* (1936) when asking "why one should write on
the subject of towns at all, when every weekend sees an ever-increasing rush
away from them into the country" (qtd. in Stevenson 193). Rouse was respond-
ing to the consequences of an alluring narrative of an idealized rural space
that had been gaining momentum in Britain at least since the end of the Great
War. A synergy of political and social factors helped to invest the countryside
with positive and desirable connotations. It came to be associated with the
power to cultivate health, console the mind, restore spiritual wellbeing, and
nurture a self-confident nation, the sense of citizenship and patriotism, which
mobilized the interwar masses, brought them to the rural space, and played a
crucial role in proliferating the patriotic spirit as the 1930s world was spiralling
towards war.

The interwar appreciation of the countryside was part of the mission to
rebuild the nation after the Great War. As Ina Zweiniger-Bargielowska illustrates, in response to the rapid urbanization and strenuous impact of the War,
the cultivation of the body was promoted by public health officers, pressure
groups, such as the Sunlight League and the New Health Society. Governments
and the National Fitness Campaign had the ambition to build an "A1 nation." "A
well-managed body," Zweiniger-Bargielowska argues, "was not only the goal of
social policy but also an integral aspect of fashioning the self" (3). Hiking in the
countryside, along with exercise, sunbathing, and other forms of "body management" were promoted as ways to build a disciplined, fit, and healthy nation
necessary for a new form of active citizenship (151–192).

The rural space was also entrusted with the capacity to cultivate personal
skills and character, an agenda that turned nature into a means of consolation and moral betterment compensating for the ills of the desolate and
demoralizing urban space. The status of nature as a superior propaedeutic
and recuperative environment acquired a collective and organized dimension. Robert Baden-Powell and Frederick Haydn Dimmock, for instance, relied
on the virtues of scouting. Having started spontaneously after Baden-Powell's

publication of *Scouting for Boys: A Handbook for Instruction in Good Citizenship* in 1908, it gained popularity as an organized movement during the interwar years. From the very beginning, Scout leaders promoted stays in nature and activities designed to cultivate the virtues of equality, self-reliance, self-sacrifice and self-discipline as prerequisites for nurturing a desirable character. In *Lessons from the Varsity of Life* (1934), Baden-Powell sums up his intent to compensate for the failure of the formal education to "inculcate a good many qualities not enunciated in school text-books" but necessary for making "reliable men" by taking young people "back to nature and backwoodsmanship, by taking the men back as nearly as possible to the primitive." His ambition was to develop in young boys "a measure of pride in their work, confidence in themselves, and a sense of responsibility and trust" through organized exposure to the natural environment with the goal of putting young people "on to a higher standard of manliness, self-respect and loyalty" (140). The objective was to use the training of character qualities in the natural environment for nurturing civic virtues: "Scouting," he claimed already in *Scouting for Boys*, "is suggested as an attractive means towards developing character and good citizenship" (295).[1] Scouting evolved alongside other nation-wide projects that contrasted the grimy urban space and its ebbing morals with the countryside viewed as a resource of a more authentic experience and as an environment with the potential to educate and cultivate the nation. For example, *New Education Fellowship*, established in 1921 to revitalize the curriculum, supported educational activities outside the classroom and the emergence of rural schools. Rural education based on the notion of the countryside as a healthy environment stimulating curiosity became a strong feature of interwar school reforms. As Alice Kirke shows, alongside the scouting movement, numerous other progressivist educational initiatives evolved with the aim to nurture a "rural community" as "a means to develop social service, citizenship and democracy" (242).

Such designs for the cultivation of the body, character, and community were accompanied with an official rhetoric of patriotism diffusing the image of the rural space as a national jewel to be cherished. In an anachronistic and nostalgic voice, Stanley Baldwin, a great admirer of Mary Webb's pastoral fiction, informed the members of the Royal Society of St. George in 1924 that the English countryside was a preserve of the true England which needed to be protected against the inflow of cosmopolitan influence. "England is the country, and the country is England," he concludes and supplies an array of sensory images supporting his claim: "the tinkle of hammer on anvil in the country

---

1 For a study of similar ambitions for camping in Germany and the USA, see Cupers, Kenny. "Governing through Nature: Camps and Youth Movements in Interwar Germany and the United States." *Cultural Geographies*, vol. 15, no. 2, April 2008, pp. 173–205.

smithy, the corncrake on a dewey morning, the sound of the scythe against the whetstone, and the sight of a plough team coming over the brow of a hill" (6–7).

Such a correlation between patriotism and idealized romantic images derived from the pre-industrial English countryside gradually strengthened and found their way to interwar intellectual debates, writing and other artforms, resulting in an almost imperative need for the public to acquaint themselves with the countryside because it formed the basis of national consciousness necessary for cultivating citizenship. Numerous artists and publishers started to partici-pate in such a mode, contributing to the growing ethos of romantic national-ism, patriotic ruralism, and "landscaped citizenship" (Matless 15). Travel books, guides, magazines and posters proliferated the notion of the countryside as a reliable, "admirable and ancestral" (Lowenthal 215) gem spared from the processes changing the face of the world. Together with a frequent empha-sis on the importance of insularity, such idealized representations of Britain's rural space assisted the official political discourse in cultivating national con-sciousness and contributed to the depiction of the countryside as an authentic national treasure worth knowing, preserving and defending. As Valentine Cun-ningham reminds us, *The Criterion*, for example, became a "house journal for the spokesmen of post-war British ruralism," with contributors such as John Betjeman or George Macaulay Trevelyan. Its chief editor, T. S. Eliot, suggested in 1931 that the majority of any society should dwell in the countryside and depend on it because the agriculture is "the foundation of the Good Life in any society; it is in fact the normal life" (*British Writers of the Thirties* 231).[2] During the 1920s, magazines such as Edward Hudson's *Country Life* or Richard Blatchford's originally socialist *Clarion* gradually came to emphasize the rural-ist agenda. They joined the interwar trend of pastoralizing the countryside and helped to establish the "industry of country writing" that tempted the urban masses into seeking the rural idyll (Howkins, "The Discovery of Rural England" 101). Numerous interwar writers also participated in the proliferation of the rhetoric of rural romanticism issuing from the tradition laid by the Georgians, Mary Webb, the collected works of which were appreciatively republished in the 1930s, and the Great War propaganda posters published by Parliamentary Recruiting Committee, such as the famous "Your Country's Call" (1915), which shows a soldier pointing at a piece of the pastoral English countryside and saying "Isn't this worth fighting for? Enlist Now" ("Your Country's Call"). John

---

2 For a detailed study of the role of agriculture in interwar debates about culture in *The Criterion*, see Diaper, Jeremy. "The 'Criterion': An Inter-war Platform for Agricultural Discus-sion." *The Agricultural History Review*, vol. 61, no. 2, 2013, pp. 282–300.

Betjeman, steering the Shell Guides, and several other writers embraced the emerging travel writing industry. Their publications, some of which directly referred to the agenda in their titles, served pastoral nostalgias, assisted the urban crowds in their weekend escapades, and helped to root in the interwar culture the idea of the rural landscape as a national treasure, source of edification and consolation. *In Search of Scotland* (1929) and *In Search of Wales* (1932) by H. V. Morton, for example, established the "motoring pastoral genre" (Matless 64) designed to guide travellers searching for the "real" and authentic in the rural parts of Scotland and Wales, respectively. In *In Search of England* (1927), he echoes interwar politicians in equating the exposure to the English countryside and intimacy with rural life – the "Back to the Land" movement – with the notion of national survival. Only those who explore and acquaint themselves with the countryside, he sums up in the introductory note, are "a step nearer that ideal national life" since "the greater the number of people with an understanding love for the villages [...] the better seems our chance of preserving and handing on to our children the monuments of the past, which is clearly a sacred duty" (VIII). Numerous other titles, such as *A Charter for Ramblers* (1934) by C. M. Joad, *Pilgrim Cottage* (1933), *Gone Rustic* (1934), and *Gone Rambling* (1935) by Cecil Roberts, provided stories, instructions, tips and advice regarding rural explorations, catering for the needs and sentiment of travellers lured by the charms of the pastoral experience.

The alluring effect of the idealized rural space was further nurtured by pastoral representations of the countryside in interwar posters. In his *Posters and their Designers* (1924), Sydney R. Jones referred to his age as "the day of the poster" (Jones VIII) and British railway companies played a crucial role in this trend. Poster anthologies by Lorna Frost, Michael Palin, Beverly Cole and Richard Durack show that railway advertising followed the dustjackets of interwar travel guides and displayed similar arcadian versions of Britain's rural landscape, transforming the woods, hedges, fields and seaside into the most common means of whetting the appetite of prospective travellers hoping for leisure and escape from the urban space.

The interwar correlation between Englishness and the countryside through pastoralized imagery continued until the Second World War, when it strengthened and nourished the patriotic sentiment. As Paul Fussell famously asserted in relation to the First World War, "if the opposite of war is peace, the opposite of experiencing moments of war is proposing moments of pastoral" (Fussell 231). In 1940, Harry Batsford published *How to See the Country* designed as a guide to help people find the "real England" in the country and learn to appreciate what is valuable for the nation. In its introductory note, he boldly sums up the relevance of his book: "[i]t is pitiable that so many English folk – possibly,

God help us, a large majority – should be so desperately out of touch with the real England. It is a state of affairs full of menace for the general welfare [...]. No one is a true Englishman, or has lived a fully-balanced life, if the country has played no part in his development" (1). In the anthology *The Englishman's Country* published in 1945, Edmund Blunden still presents this volume of essays about the English countryside as "poetry of real life in England" ("Introduction" 8). Like John Betjeman, Vera Sackville-West, Harry Roberts and other contributors to this anthology, he locates authenticity and worth in the countryside: "to the man or woman who is desirous of finding the best in this country I commend the English village" because the countryside preserves what is lost elsewhere: the "spirit of the place" in a "country community" and "native fields" surrounded by the natural landscape ("English Villages" 12–22).

These aspects of the interwar cultural climate show that a clear-cut polarization of the city and the country was becoming a widely spread trope. The natural space was infused with superior connotations deriving from its alleged power to reinvigorate the body, console the mind, and improve the morals decimated by the urban mayhem. In 1933, Cecil Day Lewis rejoiced to arrive in the genuine "country at last" after leaving behind the "barren soil" of the inauthentic, artificial, and internationalized cityscape made up of "factories, sports grounds, tame rusticity of 'garden cities', bogus Elizabethan villas, petrol stations disguised as mosques, chapels like mausoleums and amusement parks like death" (39–40). Baden-Powell too notes in *Lessons from the Varsity of Life* that stays in nature offer a wholesome refuge from the "man-made squalid surroundings of bricks and mortar" (273). Romantic mythical geography was being awakened, with the countryside seen as a paramount space offering consolation through one's immersion in an environment that is purer than the bleak and superficial city. The inauthentic and ingenuine urban space, as Cecil Day Lewis implies, provides conditions for an equally withered, shallow, and soulless existence. Can urbanites, he asks rhetorically, live on "tinned food, cheap cigarettes, votes, synthetic pearls, jazz records and standardised clothing which the town gives them back, as a 'civilised' trader gives savages beads for gold?" Their life, he concludes, is in consequence far removed from the "slow, instinctive, absorbent vision of the countryman" (40).

To many intellectuals and the general public, the countryside became such a cleansing, consoling, and regenerative panacea for the exhausting deficiencies of the period and city life (Mandler 461). This type of spatial polarization and exalted rhetoric participated in constructing what Valentine Cunningham calls the interwar "pastoral mindedness" ("Marxist Cricket?" 177) instigating mass escapism to nature. The masses were united by the notion of the rural space as an affordable, accessible, and effectual means of attaining comfort and relief, albeit temporary, from the quotidian working life in the urban space. "That was," Haydn Dimmock noted in 1937, "what Baden-Powell was offering to us – the outdoor life away from the towns" (16). The vogue for rambling and hiking as consolatory activities culminated in the 1930s, by which time the "back-to-the-land movement" had become a highly organized pastime and social force. What started as a spontaneous activity, gradually became a matter of planning and professional management. Numerous associations, such as the *National Council of Ramblers' Federations* (1932) and the *Ramblers' Association* (1935), enabled individuals keen on the exploration of nature to unite in organized groups and exercise power on authorities (e.g. through collective trespassing). The working classes were encouraged to travel and indulge in outdoor leisure activities by legal changes such as Holiday with Pay Act passed in 1938, which marked the end of a twenty-year campaign for paid leisure time. It legislated the right to leisure, and it led to the growth of the travel industry, which began to emerge hand in hand with the "Social tourism policy" that acknowledged the positive effects of travelling. Governments subsidized travel expenses, turning the charabancs and trains operated by the four main rail companies into major means of interwar mobility that transported near forty million passengers in 1939. Besides, as Victor T. C. Middleton and L. J. Lickorish add, private businesses started to provide affordable trips, excursions, and subsidized transport (1–16). The bicycle also became crucial in the interwar open-air movement and rediscovery of the rural landscape. It is estimated that there were about 10 million bicycles in Britain by the end of the 1930s. It made the countryside more accessible and, especially within the ranks of the working classes, it "freed millions from their urban confines and promoted an attitude shift toward the countryside" (Middleton and Lickorish 5). John Pimlott has argued that the appeal of the bicycle to the masses lay not only in its novelty, but also in affordability and the freedom it offered. Cycling became so popular that by 1938 the *Cyclists' Touring Club* and *National Cyclists' Union*, representing some 3,500 cycling clubs with around 60,000 members, mainly from the middle and working classes, were able to persuade local authorities to improve roads and road signs, and offer the cyclists affordable accommodation in boarding houses (Pimlott 167, 234).

Finally, numerous (volunteer) associations, such as the Youth Hostels Association, assisted hikers with cheap accommodation. Its growth was an obvious success: while in 1931 the YHA had 20 hostels and a handful of members, by 1936 over 60,000 members had been enrolled who could find accommodation in one of YHA's 260 registered hostels (Taylor 252). Besides hostels, large holiday camps, like that founded by Fletcher Dodds at Ormsby in 1906, allowed the masses to spend whole weekends or longer holiday breaks in the countryside

(Howkins, *The Death of Rural England* 26). Clearly, an effective confluence of political, social and economic factors encouraged the masses to leave the city, travel to the seaside or the countryside for the weekend or a short break and participate in strengthening the open-air vogue that came to characterize the period.[3]

The popularity of rural escapism inevitably called for the need to protect the countryside. That this became a question of a nationwide relevance is clear from various conservationist efforts signalling the onset of environmental awareness. Delayed by the Victorian emphasis on utilitarianism, Peter Mandler claims, the need to conserve the countryside as an irreplaceable component of national heritage was at last recognized by several institutions and organizations (460).[4] It is also clear from the "fortress approach" (Kim et al. 374) showing official willingness to demarcate exclusion zones. Formed in 1926, Council for Preservation of Rural England, for example, initiated legal steps, such as the demarcation of green belts, to protect the countryside from the pressures of the urban sprawl. The interwar period thus saw the diffusion of the idea of a rural landscape as a vital ingredient of Englishness because, as Patrick Abercrombie implied in 1926, "the most essential thing which *is* England, is the Countryside, the Market Town, the Village, the Hedgerow Trees, the Lanes, the Copses, the Streams and the Farmsteads." For an Englishman "[t]o destroy these," he adds, is to "lose his greatest possession – the country setting" (6).

At the same time, a voice of disapproval or even disdain with the escapist whimsy of the urban masses could be discerned in the interwar intellectual discussion. The reasons for the rejectionism varied, ranging from the preference of a solitary exposure to the countryside allowing Wordsworthian introspection, through early calls for environmental protectionism, to the critique of the rural tourism as an act of ivory-towerism and social irresponsibility. Henry Scott Holland, for example, received the attraction of the masses to the rural space with reluctance already during the Great War. In *A Bundle of Memories* (1915), he calls urban dwellers swarming in the open air "the vast, blind, welded masses." He describes them as "hordes" of "reeking babies," "dishevelled mothers," and fathers with "haggard" faces that gather at a railway station and transform it into a "wild pandemonium." Holland's contemptuous tone reflects his vision of the deleterious consequences that the arrival of the "black bunches" of uncultured masses has for the rural space. Although they come

---

3 I provide a more detailed overview of the economic, social and political context of the expanding interwar tourist industry in *W. H. Auden's Interwar Landscapes: Roots and Routes.* Routledge, 2022, pp. 1–28.

4 See also Matless, David. *Landscape and Englishness.* Reaktion Books Ltd., 1998.

with the intention of immersing themselves in "the deep bosom of Nature, to steep [themselves] in her calm, to feed on her changeless and eternal peace," Holland thinks that the crowds fail to encounter such states. Conditioning the attainment of a profound consolatory communion with the natural world by instruction and refinement, what he thinks is the main obstacle for the uncouth crowd is their materialist and consumerist predisposition. They arrive with "lunch-baskets and bottled beer" to feast and make noise, consequently failing to attain consolation through a spiritual rapport with nature. To "drink of Nature's draughts," he proposes, one needs to be alone and patient. Nature can induce consolatory experience but it "only speaks her full mind to those who can be alone with her: and she cannot be hurried. Her speech is slow-distilled." The deserving traveller also needs to possess the virtue of patience. "It takes time to lean an ear to fairy-waterbreaks," Holland proposes, "if you really desire the beauty born of murmuring sound to pass into your face. This is not the sort of thing that can be managed in the interval while the rowdy coach is waiting." Holland's views are exclusivist as only the deserving ones – the slow, ponderous, and solitary travellers – can benefit from the consoling effects of the natural environment. Crowds and those who rush lose "the sense of 'winds austere and pure' blowing over 'grey recumbent stones in desert places on the naked wine-dark moor'." Here, Holland alludes to R. L. Stevenson's famous poem "To S. R. Crockett" (1895), but in fostering the restorative, mystical, and inspiring potential of the natural world through solitariness and ponderous contemplation, his mindset is Wordsworthian, which he finally admits: "whenever the trip comes in, Wordsworth goes out" (241–242).

Unlike Holland, George Macaulay Trevelyan combined the belief in nature as a consoling place with preservationist concerns. Aware of the budding public fondness for the countryside and the need to organize the masses in their explorations in order to protect the landscape from spoliation, in 1931 Trevelyan became the first president of Youth Hostels Association, the mission of which was "to help all, especially young people of limited means, to a greater knowledge, love and care of the countryside [...] and thus to promote their health, recreation and education" ("YHA – England and Wales"). He also purchased land, donated it to the National Trust, and joined the organization to help institutionalize conservationist processes because, as Rick Rylance summarizes, his charitable and conservationist involvement was animated by concerns about the "destruction of the countryside and its heritage" and "the loss of natural beauty" (87). Trevelyan's preservationist agenda evolved alongside his growing faith in the positive consolatory effects of immersion in nature. Convinced of the curative potential of country walking, he presaged the body management trend and the belief in the restorative power of the rural space that were to unfold between the Wars. "I have two doctors, my left leg and my right," he says in the opening sentence of his essay *Walking* (1913). "When body and mind are out of gear [...] I know that I have only to call in my doctors and I shall be well again" (19). The essay provides readers with more than mere tips on the types of roads to be taken and appropriate ways of communicating with the locals. It goes beyond such practical aims of Morton's *In Search of England*, Batsford's *How to See the Country* and other interwar publications referred to earlier in that it offers a self-assured pedagogy and philosophy of immersion in nature. Trevelyan claims that a slow and ponderous movement through the rural space not only helps the body recuperate, but it also delivers coveted consolation and spiritual experience derived from a profound state of communion with nature. Once in the countryside, the urban traveller feels "the higher ecstasies of Walking [...] through the whole being" and is rewarded with "the repossession of his own soul." Reminiscent of Holland, Trevelyan prefers quiet walking, the positive effects of which are unavailable to the crowds. Only in "silence and solitude" can "the harmony of body, mind, and soul" occur because they are "no longer conscious of their separate, jarring entities." Consoling oneness can be experienced in precisely this moment, allowing "mystic union with the earth, with the hills that still beckon, with the sunset that still shows the tufted moor under foot" (20–26). Trevelyan clearly expresses some of the central ideas that would characterize the interwar approach to the countryside. In his writing he treats the rural space not only as a crucial component of national heritage in need of protection. He also conceives of it in a strictly anthropocentric way as a source of a rare experience of unity and harmony with the Earth, which comforts the body and mind fatigued by the urban space.

Mass escapism and fascination with the countryside were also of prime concern for W. H. Auden. In the interwar period he earned the reputation of an urban poet with a fondness for industrial imagery and a topophilic sentiment for former lead-mining sites of the North Pennines. Yet the world of nature represents one of the most crucial focal points of Auden's imagination and ruminations. In its treatment, however, he clearly dissents from the period trope of nature as a recuperative refuge and consoling place. In reflection of his well-recorded indebtedness to Thomas Hardy, frost, wind, and gale in his poems kill animals and humans, rendering nature a cruel, merciless, and indifferent force. While such unromantic images abound in his work, especially in the early years of his career, the staples of Auden's poetics of nature are awe, respect, and admiration for the natural environment as a superior realm of contentment. As shown by numerous critics (e.g. Mendelson 1999; Fuller 1998), the vitality of life in nature is rejuvenated by ceaseless cyclical processes and energy that springs out of the Earth to guarantee thriving and prosperity. Every morning, the cock "summon[s] up/The pointed crocus top" (*Poems I* 60), "magnificent clouds" "move without anxiety" while "a colony of duck below/Sit, preen, and doze on buttresses/Or upright paddle on a flickering stream/Casually fishing at a passing straw." All components of the natural world live at rest, finding "sun's luxury enough" (*Poems I* 37). In *Letters from Iceland* (1937) too, Auden is surgically precise to craft lyrical images conveying the purity of the natural world. The island abounds in the "glitter of glaciers," "sterile immature mountains," birds that "flicker and flaunt," and other "natural marvels" (*Prose I* 185) that render Iceland a sanctuary of a pristine preindustrial rural landscape. In much later poems, Auden's awe persists. If "our Mother" could look into a mirror, Auden pays homage to the Earth in "Ode to Gaea" (1954), she would see herself "Far-shining in excellence" (*Poems II* 427).

While such nature lyricism weaves through Auden's oeuvre, the poet never embraced the idea of entering the natural space with consolatory, recupera-tive, or cultivating aspirations. What Auden shared with Holland, Trevelyan, and other intellectuals of the time was a grudge against the open-air move-ment and the masses seeking comfort in a rural retreat from the city. Yet his reasons were neither protectionist concerns nor an anticipation of the neg-ative impact of crowds on the attainability of a mystical unity with Mother Nature. Unlike John Pimlott, who looked back on the interwar years and sym-pathized with the "imprisoned millions" of interwar urban workers in need of "little persuasion to escape when they could" (213), and unlike Holland and Trevelyan, who called for a more ponderous and inspiriting approach, Auden thought of rural pursuits in terms of weakness and irresponsible isolationism caused by a manipulative spatial myth. As I have shown elsewhere (Vít 2022), Auden's criticisms of the open-air movement and the lure of consolatory com-munion with nature were not marks of his contempt for the natural world, which he admired. His critique was firmly rooted in his ethical commitment to the urban space and in his down-to-earth sobriety which was in part derived from the scientific leanings of his family: Auden's brother John Bicknell Auden studied geology at Cambridge before working as a land surveyor, their father George Augustus Auden maintained a keen interest in geology and mining, and Wystan himself came to Oxford to study natural sciences.

In Auden's writing, any sign of escapism or hankering for a rural retreat and spiritual experience of unity with nature are scorned as symptoms of an ego-istic evasion of civic duties. This interwar trend, Auden insisted, is nourished by an uncritical submission to a romantic view of nature as a motherly womb to which to return for consolation and unitary experience. It is clear from Auden's satirical tone that the open-air movement, "pastoral mindedness," and romantic pedestrianism started to agitate him strongly, especially when signs of an oncoming war began to appear. In his 1930s writing, he mocks the "manoeuvres of the week-end hikers" and their "passion for the open air and shorts" (*Prose I* 210). He thinks the rustic image of urban travellers as well as their desire to derive consolation from nature and a simple rural life are ingenuine. He ridicules the ingenuity of a consciously ruggish, rustic appearance and uncomfortable travel style of anyone that "doesn't shave his chin," "wears a very pretty little boot," and "chooses the least comfortable inn." He treats with contempt those who think that people staying in town are weak "wenmen" because the "Good Life is confined above the snow-line" (*Prose I* 251). As I have shown earlier (Vít 2022), Auden also uses the notion of artificiality in his criticism of Scout leaders and their objective of cultivating the citizenship and character of young people in the embrace of the natural world. Reviewing Baden-Powell's *Lessons from the Varsity of Life* in 1934, Auden argues that for the Scout leader to say that "the Backwoods life is natural and City life artificial is nonsense" because, with the majority of people in Britain living in cities, "[c]amping is really a highly artificial training for a better town life, and, valuable as it is, town life demands much more" (*Prose I* 63). A similar dissenting tone characterizes Auden's review of *Bare Knee Days* (1937) by Frederick Haydn Dimmock, a long-time editor of the magazine *Scout* and author of numerous junior scout novels. In the 1930s, Auden was a practicing teacher with clear opinions on the flaws of formal education in Britain. Hence, in the review Auden lauds Haydn's intent to teach children to be "observant and self-reliant" but rejects his spatial differentiation and claim that such goals are more easily achievable in the "real" natural environment than in the "artificial" urban space (*Prose I* 424–425). Auden's admiration of the superior natural world and his ability to write nature lyricism were matched neither with an appreciation of mass escapism to nature nor with the belief in its propaedeutic and consolatory potential.

Auden's disapproving attitude towards the interwar "pastoral mindedness" and his insistence on an unromanticized view of nature reflect not only the scientific leanings of his family. Auden's work also shows the extent to which he appropriated some of the major tenets of what would be described as the Human Exceptionalism Paradigm. As a conceptual framework, it entrusts humans with a specific type of existence, the uniqueness of which issues from an unmatched use of reason, logic, and symbolic languages (Catton and Dunlap 41–49). Human exceptionalism promotes a disconnected position of humans *vis a vis* the biophysical environment. Deriving from Cartesian dualism, the position reflects the belief that "humans and human societies exist independently of the ecosystems in which they are embedded, thereby promoting a sharp ontological boundary between humans and the rest of the natural world" (Kim et al. 358). Although anthropocentrism and human exceptionalism over-lap in their focus on the human relation to the biophysical environment, there is also a significant distinction in questions concerning our connectedness to nature. The former "overestimat[es] the centrality of humans in the natural world" and approaches "humans as central exemplars of the biological world, regardless of whether they are part of it." The latter, on the other hand, insists on the idea of exclusion and the existence of an irremovable chasm, "a hard ontological boundary between humans and other living things" (Kim et al. 362–363).

Auden championed the human exceptionalist sense of humans as "onto-logically unique and biologically discontinuous with the rest of the living world" (Betz and Coley 2), hence barred from the attainment of a profound state of consolatory communion with nature. Asking if Auden's poems can be read as ecological texts, Rainer Emig concludes that Auden did indeed place an irrevocable ontological divide between humans and the natural world. Emig illustrates how Auden used nature anthropocentrically as "a symbolic point of orientation" because he believed that "our representations of nature" are "inevitably anthropocentric" (212–215). Auden interpreted the rift in clear Car-tesian and human-exceptionalist terms. One of the central motifs in his Berlin diary (1929) is the dualism of mind and body. Like Serenella Iovino (2012) and other scholars in the ecocritical fold who stress our shared materiality with the non-human world, Auden treats the body as part of nature because it is what people share with other organisms; it is "communistic" (i.e. of commu-nity) in the sense that it tends towards sameness since the physical differences between individuals are negligible (*The English Auden* 297–298). Auden relates this likeness to his claim that the body is no longer prone to evolution because, as he would add in 1938, "[m]an finished his biological development long ago" (*Prose I* 472). He contrasts the subject-to-nature body with the properties of the mind, which, he claims, tends towards individuation. The use of the mind guarantees human uniqueness among other species, provides ground for mak-ing further progress, but it also causes our irrevocable distancing from nature. "[T]he development of the mind," Auden concludes in his diary, "is one more and more of differentiation, individualistic, away from nature." Auden found the main cause of this dissociation in self-consciousness, which he believed had created an uncrossable rift between humans and the natural world. "The development of consciousness," he notes in his diary, "may be compared with the breaking away of the child from the Oedipus relation" because in the same way as "one must be weaned from one's mother, one must be weaned from the Earth Mother." The imperative "must" shows the scale of Auden's conviction about the inevitability of human exceptionalism. This fuelled his grudge against any agenda like that proposed by Trevelyan and Holland which promised to bridge the rift between humans and nature for the sake of regaining unity and finding consolation. For Auden, such ambitions are clear signs of a futile "nostalgia for the womb of Nature which cannot be re-entered" by man compelled by self-consciousness to face the consequences of such irrevocable parting (*The English Auden* 298).

Auden's early poems can be read as laments over the unbridgeable chasm between nature and man, which turns humans into existential outcasts unable to re-enter and reconnect to the realm of contentment governing nature. However, unlike numerous other intellectuals who located the remedy in a profound state of communion with nature, Auden in the 1930s found consolation in the very opposite. He came to conceptualize the human mind as a unique evolutionary gift to be used on our journey to freedom and away from nature, albeit one paid for with the necessity of choice-making. Human exceptionalism presupposes that humans are unique because they are the only organism possessing the capacity of abstract reasoning (Kim et al. 360) and the potential to manifest free will for the sake of initiating changes to the current state of society. Reviewing Liddell Hart's biography of the soldier and Arabist T. E. Lawrence in 1934, Auden nods to the same: "self-consciousness is an asset, in fact the only friend of our progress" (*Prose I* 61). Unlike his early poems that can be read as lamentations over the losses caused by human self-consciousness and reason, in the 1930s he started to treat them as unique phenomena with the power to facilitate human progress. Reason gradually became a practical "instrument" (*Prose I* 6) which, although it could at times be "an uncertain guide," flared up to become "the light by which man must live" (*Prose I* 141).

Auden's interwar writing shows a clear development in his views on the relation between reason, free will, and change. His early poems display the human will and rational choices crushed by indifferent natural forces. Through the 1930s, as the radical politics were on the rise, Auden began to emphasize the need to use reason and will for initiating changes and conscious sociocultural evolution that would extricate man from the controlling forces of nature and manipulative systems of thought. Enumerating the main points of Freud's teaching, for example, he suggested in 1935 that, due to the evolving and individuating mind, "[m]an differs from the rest of the organic world in that his development is unfinished" (*Prose I* 101). That Auden himself accepted this tenet of human exceptionalism is clear from his other essays and reviews from the mid-1930s. In 1937, he drafted an introduction to the *Oxford Book of Light Verse*, where he claims that while life in preindustrial communities was unfree and the virtues were "nursed unconsciously by the forces of nature," the societies must now be "fostered by a deliberate effort of the will and the intelligence. In the future, societies will not grow of themselves. They will either be made consciously or decay" (*Prose I* 436). A few months after publishing the introduction, he added that man is "the only animal who has been able to continue evolution after biological development is finished" because he is "the only animal capable of using his intelligence and making choices; the only animal whose society can develop from one form into another" (*Prose I* 463).[5] Auden thus voices a crucial human-exceptionalist presupposition: "[h]umans are unique among the earth's creatures, for they have culture," which "can vary almost infinitely and can change much more rapidly than biological traits" of the human body and other natural organisms (Catton and Dunlap 42).

Critics of human exceptionalism argue that the ideas of an evolving culture and change have led to the assumption that humans are exempt from the constraints of the biophysical environment and that "biological laws do not, or no longer, apply to humans" (Betz and Coley 2). From this perspective, human "progress can continue without limit, making all social problems ultimately soluble," which has contributed to fostering the "prevalence of the doctrine of progress in Western culture" (Catton and Dunlap 43). In his prose and poetry, Auden seems to delineate precisely these contours of our being in the world – an existence of constant becoming caused by self-consciousness, reason, and culture. For example, before depriving poetry of its social potency in 1939, Auden viewed art as a cultural practice contributing to changes and progress because of its capacity to "deepen understanding, to enlarge sympathy, to strengthen the will to action and, last but not least, to entertain," which gave art "an honourable function in any proper community" (*Prose I* 134).

Ruminating on the mutual relationship between man and the natural environment, Auden provocatively found consolation not in a return to nature but a departure from it. He maintained awe and admiration for the natural world. Yet his trust in the power of reason and human evolutionary imperative to overcome unfreedom, together with his vision of a conscious path taken towards a desirable form of society, fuelled his animosity towards interwar masses charmed by the narrative of comforting immersions in nature. Agendas encouraging seclusion, instruction, consolation, and communion with the natural world agitated him profoundly. In the already cited review of Dimmock's book on scouting, Auden rejects the notion of an escape to nature as an alternative to what Dimmock calls the "artificial" urban space. Auden insists on the importance of learning to live in town because our inevitable separation

---

5   See also Auden's essay "Morality in an Age of Change" (1938) in the same volume.

from nature turns the urban space into the human habitat proper. It is in this part of the agenda, Auden rebukes Dimmock, "that scouting fails" (*Prose I* 63). Because "[m]ost of us have to live in towns," Auden justifies his claim, we "need to be taught how. All civilised life is artificial; i.e. it is life which is not ruled by the forces of Nature, but by one's own free will, and no one can use that properly who has not acquired self-knowledge and the habit of reason" (*Prose I* 425). For the same insistence on the insurmountable ontological rift between humans and nature, Auden mocked any programme promising the positive effects of their organic reunion. Calling William Wordsworth an "old potato" (*Prose I* 251), he blamed the Romantic poet for inventing the myth of nature as a place of consolation that flourished in the interwar years and encouraged the masses to escape from the cities. "Along with the growing self-consciousness of man during the last 150 years," he sums up the legacy of the Lake Poet, "has developed Wordsworthian nature-worship" (*The English Auden* 298). Auden's conviction that Wordsworth invented the persisting romanticized view of nature was his life-long grudge against the Romantic poet. It is clear, for example, from "Mountains" (1952), the third part of *Bucolics*, where Auden states his allegiance to the urban space while pointing out the false nature of the rural myth: "A civil man is a citizen. Am I/To see in the Lake District, then,/Another bourgeois invention like the piano?" (*Poems II* 409). In all his writing Auden approaches rural pursuits and romanticization of the natural world as consequences of such a worshipful but futile "nostalgia for the womb of Nature" (*The English Auden* 298) because our *sui generis* existence is irredeemably removed from the natural world. "Hell is," Auden in *New Year Letter* (1940) explains on behalf of all humanity, "if we deny/The laws of consciousness," and if we claim that "Becoming and Being are the same," and ignore the fact that each human being is "locked" in their "stale uniqueness" (*Poems II* 30).

The extent and spectrum of responses to the open-air movement show how relevant the question of (re)turning to nature for consolation was for the cultural climate of the period. A complex and conflicting moral geography of nature was fashioned. The rural space was related to themes, concerns, and agendas concerning individuals as much as the whole nation. In writing and political and institutional discourses, nature re-emerged as a more genuine type of environment offering consolation, comfort, and a therapeutic alternative to the artificial and demoralizing urban space. Nature was entrusted with the potential to assist humans in the recuperation of their bodies and minds which had been exhausted by the city. It gained the virtuous power to console the modern man and induce mystical experience in which one could attain a desirable recuperative unity with the Earth, a type of an encounter impossible in the urban space. The excursions of the masses to it, however, also triggered a variety of concerns about the morality of rural tourism and escapism from the existential urban space. To some intellectuals, seclusion in nature became a sign of an irresponsible isolation and ambition, fuelled by a romanticized spatial myth, to evade the commitment to the social space and indulge in a vain quest for seclusion and a regenerative communion with nature. The diversity of the intellectual responses shows how the ideas of nature and its power to provide consolation grew to become not only crucial components of Britain's interwar cultural practice, but also fuel for intellectual debates of an impressive variety ranging from health and spiritual wellbeing, through nationhood and citizenship, to a uniquely human responsibility for the existential space.

Bibliography

Abercrombie, Patrick. "The Preservation of Rural England." *The Town Planning Review*, vol. 12, no. 1, 1926, pp. 5–56.

Auden, Wystan Hugh. *The Complete Works of W. H. Auden: Poems, vol. I: 1927–1939*, edited by Edward Mendelson, Princeton University Press, 2022.

Auden, Wystan Hugh. *The Complete Works of W. H. Auden: Poems, vol. II: 1940–1973*, edited by Edward Mendelson, Princeton University Press, 2022.

Auden, Wystan Hugh. *The Complete Works of W. H. Auden: Prose and Travel Books in Prose and Verse, vol. I: 1926–1938*, edited by Edward Mendelson, Princeton University Press, 1996.

Auden, Wystan Hugh. *The Complete Works of W. H. Auden: Prose, vol. III: 1949–1955*, edited by Edward Mendelson, Princeton University Press, 2008.

Auden, Wystan Hugh. *The English Auden: Poems, Essays and Dramatic Writings, 1927–1939*, edited by Edward Mendelson, Faber and Faber, 1988.

Baden-Powell, Robert. *Lessons from the Varsity of Life*. C. Arthur Pearson, 1934.

Baden-Powell, Robert. *Scouting for Boys: A Handbook for Instruction in Good Citizenship*. 1908. Oxford University Press, 2005.

Baldwin, Stanley. *On England, And Other Addresses*. Philip Allan, 1926.

Batsford, Harry. *How to See the Country*. B. T. Batsford, 1940.

Betz, Nicole, and J. D. Coley. "Exceptionalist Thinking about Climate Change." *Sustainability*, vol. 14, 2022, pp. 1–28.

Blunden, Edmund. "English Villages." *The Englishman's Country*, edited by W. J. Turner, Collins, 1945, pp. 11–52.

Blunden, Edmund. "Introduction." *The Englishman's Country*, edited by W. J. Turner, Collins, 1945, pp. 5–8.

Catton, William R., and Riley E. Dunlap. "Environmental Sociology: A New Paradigm." *The American Sociologist*, vol. 13, no. 1, 1978, pp. 41–49.

Cole, Beverley, and Richard Durack. *Railway Posters 1923–1947*. Laurence King, 1992.

Cunningham, Valentine. *British Writers of the Thirties*. 1988. Oxford University Press, 1989.

Cunningham, Valentine. "Marxist Cricket? Some Versions of Pastoral in the Poetry of the Thirties." *Ecology and the Literature of the British Left: The Red and the Green*, edited by John Rignall and H. Gustav Klaus, Routledge, 2012, pp. 177–191.

Cupers, Kenny. "Governing through Nature: Camps and Youth Movements in Inter-war Germany and the United States." *Cultural Geographies*, vol. 15, no. 2, 2008, pp. 173–205.

Day Lewis, Cecil. "Letter to a Young Revolutionary." *New Country: Prose and Poetry by the Authors of New Signatures*, edited by Michael Roberts, Hogarth Press, 1933, pp. 25–42.

Diaper, Jeremy. "The 'Criterion': An Inter-war Platform for Agricultural Discussion." *The Agricultural History* Review, vol. 61, no. 2, 2013, pp. 282–300.

Dimmock, Frederick Haydn. *Bare Knee Days*. Boriswood, 1937.

Eliot, T. S. "Commentary." *The Criterion*, vol. 11, October 1931, p. 72.

Emig, Rainer. "Auden and Ecology." *The Cambridge Companion to W. H. Auden*, edited by Stan Smith, Cambridge University Press, 2004, pp. 212–225.

Frost, Lorna. *Railway Posters*. Shire Publications, 2013.

Fuller, John. *W. H. Auden: A Commentary*. Faber and Faber, 1998.

Fussell, Paul. *The Great War and Modern Memory*. Oxford University Press, 1975.

Holland, Henry Scott. *A Bundle of Memories*. Wells Gardner, Darton & Co., Ltd., 1915.

Howkins, Alun. *The Death of Rural England: A Social History of the Countryside since 1900*. Routledge, 2003.

Howkins, Alun. "The Discovery of Rural England." *Englishness: Politics and Culture 1880– 1920*, edited by Robert Colls and Philip Dodd, Bloomsbury, 1987, pp. 85–112.

Iovino, Serenella. "Material Ecocriticism: Matter, Text, and Posthuman Ethics." *Literature, Ecology, Ethics: Recent Trends in Ecocriticism*, edited by Timo Müller and Michael Sauter, Universitätsverlag Winter, 2012, pp. 51–68.

Jones, Sidney R. *Posters and their Designers*. The Studio, 1924.

Kim, Joan J. H., Nicole Betz, Brian Helmuth, and John D. Coley. "Conceptualizing Human–Nature Relationships: Implications of Human Exceptionalist Think-ing for Sustainability and Conservation." *Topics in Cognitive Science*, vol. 15, 2023, pp. 357–387.

Kirke, Alice. *Education in Interwar Rural England: Community, Schooling, and Voluntarism*. 2016. University College London, PhD dissertation.

Lowenthal, David. "British National Identity and the English Landscape." *Rural History*, vol. 2, 1991, pp. 205–230.

Mandler, Peter. "Politics and the English Landscape since the First World War." *Huntington Library Quarterly*, vol. 55, no. 3, 1992, pp. 459–476.

Matless, David. *Landscape and Englishness*. Reaktion Books Ltd., 1998.

Mendelson, Edward. *Early Auden*. 1981. Faber and Faber, 1999.

Middleton, Victor T. C., and L. J. Lickorish. *British Tourism: The Remarkable Story of Growth*. Butterworth-Heinemann, 2007.

Morton, Henry V. *In Search of England*. 1927. Methuen and Co. Ltd., 1937.

Palin, Michael. *The Golden Age of the Railway Posters*. Pavilion Books Limited, 1987.

Pimlott, John, A. R. *The Englishman's Holiday: A Social History*. Faber and Faber, 1947.

Rylance, Rick. "Three North Pennine Journeys in the 1930s: Auden, Trevelyan and Wainwright." *Critical Survey*, vol. 10, no. 3, 1998, pp. 83–94.

Stevenson, John. "The Countryside, Planning, and Civil Society in Britain, 1926–1947." *Civil Society in British History: Ideas, Identities, Institutions*, edited by Jose Harris, Oxford University Press, 2003, pp. 191–211.

Taylor, Harvey. *A Claim on the Countryside: A History of the British Outdoor Movement*. Edinburgh University Press, 1997.

Trevelyan, George Macaulay. *Walking*. 1913. Edwin Valentine Mitchell, 1928.

Vít, Ladislav. *The Landscapes of W. H. Auden's Interwar Poetry: Roots and Routes*. Routledge, 2022.

"YHA – England and Wales." *Charity Commission for England and Wales*, 19 November 2023, https://register-of-charities.charitycommission.gov.uk/charity-search/-/charity-details/306122. Accessed 20 Feb. 2025.

"Your Country's Call. Isn't this worth fighting for? Enlist Now." Poster. Jowett and Sowry, 1915. https://collection.nam.ac.uk/detail.php?acc=1977-06-81-22. Accessed 20 Feb. 2025.

Zweiniger-Bargielowska, Ina. *Managing the Body: Beauty, Health and Fitness in Britain, 1880–1939*. Oxford University Press, 2010.

CHAPTER 4

Reclaiming Nature: John Hargrave's Kibbo Kift and the Vision of Post-World War I Renewal

Izabela Curyłło-Klag

Abstract

This chapter explores the connection between woodcraft and modernism in 1920s Britain to unveil a post-World War I utopian shift that grounded its aesthetic intervention in nature. The primary focus of my analysis is the initial, apolitical phase of John Hargrave's Kibbo Kift Kindred, characterized by a unique creative culture, rooted in an appreciation of the natural world. Seeking inspiration in a vast pool of "rejected knowledge" of alternative sciences and beliefs of the previous epochs, Hargrave proposed a distinctive programme of moral and physical regeneration in order to break the cycle of industrial growth, and to counteract the "unnatural" forces that were propelling the world towards destruction. Advocating outdoor activities as a remedy for the adverse impacts of industrialism and war, the Kibbo Kift aimed to reach global youth, fostering a shared vision for the peaceful coexistence of nations. Simultaneously, the movement sought to engage contemporary artists and intellectuals in a grand scheme of social renewal. After the upheaval of 1914–1918, a return to the outdoors was deemed imperative, rooted in the belief that only in nature's restorative and consolatory powers could humanity heal, regain its vitality, and truly flourish. This initiative resonated with some strands of modernism which sought alternative paths toward harmony and unity in the aftermath of a war-torn world. The chapter challenges the recent revisionary trend in modernist studies, which often portrays the modernist movement as endorsing the destructive facets of modern civilization. Instead, it illuminates an emerging green countercultural discourse within modernism, generating pockets of resistance against prevailing forms of power. At the same time, attention is drawn to the more disquieting aspect of the modernist ambition to breed a new elite of world leaders ready to intervene in the case of a global emergency.

Keywords

return to nature – woodcraft – modernism – social resistance – societal healing

In a recent book, *Utopianism for a Dying Planet: Life After Consumerism* (2022), Gregory Claeys calls for a reconsideration of the utopian tradition, arguing that it should not be dismissed as belonging to the realm of the impossible but rather, it may actually light a path to a more sustainable, fairer, ecologically-aware society. He points out that "utopians have been found at the forefront of numerous progressive movements for five hundred years" (15), and postulates reviving utopianism's humanitarian agenda so that it is not viewed just as a way of escaping reality but a powerful means to change it. Building upon this premise, this chapter will focus on the modernist utopia envisioned by John Hargrave and the Kibbo Kift Kindred, who, in the aftermath of World War I, endeavoured to implement a scheme of social regeneration through creativity and appreciation of nature. In contrast to many figures of mainstream modern-ism who enthused about the city, Hargrave drew attention to the countryside as a curative, restorative space, idealizing it not as a place of employment but as one of anarchic play and liberation. He believed that engaging with the natural world contained the remedy for the scars of the war and the promise of human-ity's improvement through fostering an international network of like-minded, ecologically-conscious, and pacifist individuals. His activism foreshadowed many later countercultural calls for environmental sustainability in Britain, and also served as a pocket of resistance within modernism in that it postu-lated a departure from the industrial and urban focus of the time. Although relatively small-scale and existing on the fringe of avant-garde cultural produc-tion, Hargrave's unfinished project is an important example of what Jeffrey Mathes McCarthy classifies as "green modernism,"[1] combining artistic creation with civic concern to challenge the dominant anthropocentric narrative and the nature-culture binary.

A split-off from the Scouts, the Kibbo Kift[2] Kindred came into being in August 1920, at a time when "radical change was called for and radical experiments were welcomed" (Pollen 11). The group's ambitions were impressive from the outset: to engage youth globally and encourage outdoor activities, counteracting the harm-ful effects of mechanization and war. Woodcraft[3] and living in the elements were not only intended to reinvigorate the postwar generation but also to provide

---

1   See J. M. McCarthy, "'The Land's Way Is Important in This Story': Environmental Criticism in Modernist Studies" in: *Green Modernism: Nature and the English Novel, 1900–1930* (Palgrave Macmillan, 2015), pp. 1–40.
2   Kibbo Kift means "proof of great strength" in the Cheshire dialect.
3   Inspired by the ideas of a Canadian ranger, Ernest Thompson Seton, woodcraft was a method of outdoor training, based on an application of primitive and ancient skills, that were meant to reinvigorate the practitioner, both physically and spiritually. It arose from the fascination with the Native American ability to survive in the wilderness.

solace and healing through nature's restorative power, laying the groundwork for a new humanity: fit, enlightened, and able to coexist in peace. The creation of the League of Nations must have been an important inspiration: in February 1920, its Council held a meeting in London, and most of the proceedings were open to the public and widely commented upon in the press. The organization's treaty and the manner of calling assemblies gave Hargrave the idea of how to run an international movement. The Covenant signed by Kibbo Kift members echoed Woodrow Wilson's program for world peace, advocating international freedom of trade, open negotiation, disarmament of all nations, and the establishment of a World Council. Yet the most important point, specified at the beginning of Hargrave's list of objectives, was:

> [t]o counteract the ill-effects of industrialisation and overcrowding in towns and cities by giving the rising generation the opportunity of a full education (in which Camp Life is an important factor), by establishing Woodcraft Groups, Camp Schools, Open-Air Schools, Holiday Camps, Local Camping Grounds, Land reservations and Open Spaces for camp training and Nature-craft. (*The Mark*, June 1922, 16)

In contrast to the Boy Scouts, which Hargrave had belonged to since 1908, the Kibbo Kift was primarily conceived in reaction to the war and the rapid changes transforming postwar society. Although Hargrave achieved a high position in Baden-Powell's organization, he was repelled by its increasingly militaristic training. Some ideas were shared and continued: woodcraft and camping as a way of life, developing self-discipline through wilderness survival, fostering resilience. Yet ideologically, the focus had shifted: rather than civilizing boys for imperialist social ends, Hargrave laid emphasis on shaping character through communion with nature and learning to conserve and protect it. Collective and coeducational, his camps were meant as experiments in the simplification of life, an alternative space where moral re-education was possible, away from the corruptions of the city, wants and prejudices of conventional lifestyle. Having formed his splinter group, Hargrave gave interviews in the press where he was described as "a firm believer in nature and in natural methods of training the youth of the country," opposing military discipline and arguing that "if people were brought more in contact with nature as children, they would be less likely to run amok, and as a nation, get up those ridiculous 'hates' of which they heard so much" (*Huddersfield Daily Examiner* 2).

With his Quaker background and experience as a stretcher bearer in the Gallipoli campaign, Hargrave was sceptical of all war-mongering rhetoric. He consistently emphasized his movement's pacifism and was careful, at least in its initial phase, to steer clear of both left- and right-wing ideologies. He ensured that the Kindred were represented in No More War demonstrations in Hyde Park. In an early letter to *The Mark*, the Kibbo Kift press organ, he made his position clear:

PEACE BEGETS PEACE

If you want peace, prepare for peace, but whoso [sic] prepares for war (by white armies or red armies) shall reap death, famine, pestilence and terror, and the result shall be more war – and yet another war – followed by still others without end.

Those who are members of the Kibbo Kift cannot have a hand in this war game whatsoever. Others may wave aloft Union Jacks or red Flags, but for us it is the World Flag of the Kindred. (*The Mark*, September 1922, 55)

Hargrave's diverse creative talents enabled him to effectively express his anti-war stance. He regularly produced protest graphics and leaflets for pacifist events. His remarkable series of banners for London's 1926 Peacemaker's Pilgrimage featured slogans such as "War is Murder," "War is Hell," "War Won't Work," and "Life or Death" – the latter depicting a nature-worshipping youth contrasted with a robotic soldier figure wearing a gas mask. Additionally, the Kibbo Kift leader was a published author and illustrator of his own texts. His 1919 anti-establishment treatise, *The Great War Brings It Home: The Natural Reconstruction of an Unnatural Existence*, spelled out its chief message already in the title. It was, as Mark Drakeford describes it, "a vast, unsynthesised outpouring of ideas and proposals for the reform of society," as well as "one of the first attempts in England to delineate a response to [wartime experience] and imbue it with a new meaning" (37).

Spanning over 360 pages, the book criticized Western culture for producing the catastrophe of global conflict and sending the prime of its youth into the trenches. In language influenced by eugenics, Hargrave lamented the military's consumption of able-bodied individuals, leading to a demographic imbalance that threatened the vitality of post-war generations. Depleted and exhausted, humanity now needed to "progress backwards" and revert to more organic, healthy ways of living. However, Hargrave had little faith in society's leaders to engage in critical reflection. He complained, "Most of the plans suggested for reconstruction after the war in education, industry, and politics are totally useless because the rulers have not the courage to abandon the mechanized civilized slavery which by an unseen course brought about the war" (*Great War Brings It Home* 360), and instead proposed to resolve the problem at the grassroots level.

Hargrave further elaborated on this idea by arguing that humanity's gravest mistake was its abandonment of the ancient heritage of hardihood in favour of a civilized lifestyle that ultimately propelled it towards self-annihilation:

> The wild, hardy life is natural and permanent. The other temporary and artificial. To get 'out of touch with nature' is to get out of touch with life itself and yet, our modern civilisation did its best to ignore nature in every possible way. Such a civilisation tends to destroy life and the Great War brings it home. (*Great War Brings It Home* XVI)

Hargrave contended that weak civilizations are inevitably swept away by hordes of barbarians, and thus argued that the contemporary world needed to cultivate a new "savage stock," who would embark on a great crusade of hiking and camping in the name of health, world government, and international peace. These "Intellectual Savages," as he called them, would be the bearers of the new gospel of a return to nature, informed by vitalist scientific thinking and infused with elements of mysticism.

Hargrave's perception of his mission in spiritual terms became evident when he tried his hand as an author of fiction in 1924 and published *Harbottle: A Modern Pilgrim's Progress from This World to That Which Is to Come*. The novel, an allusion to John Bunyan's theological narrative, follows the story of an Everyman figure, John Christian Harbottle, as he grapples with the profound disruptions caused by the Great War. Harbottle's personal losses are acute: his younger son perishes in combat, while his older son returns home maimed by a landmine, only to die subsequently. Harbottle himself is called up for service shortly before the war's end. During a period of home leave, he finds his domestic life in ruins – his wife has left him for a young major. To salvage his sanity, he opts for the life of the open trail and derives comfort from the encounters with the natural. The consolatory power of the environment becomes the narrative's central theme, offering solace and a sense of renewal amidst the devastation.

Central to Hargrave's vision was a utopian belief in the necessity of self-recovery – physical, intellectual and spiritual – as the foundation for broader societal healing. The "tribal training" he proposed for the young was meant to instil a reverence for nature and cultivate a sense of wonder, preparing new generations to live in harmony with the environment and one another. The symbol of the movement, the letter "K" suggested a human being standing upright in worship, one hand stretching towards the earth, the other towards the sky. Drawing inspiration from a vast pool of "rejected knowledge" of alternative beliefs from previous epochs (e.g. the philosophy of Thoreau and

Carpenter, New Romanticism, New Age mysticism, the Occult Revival), the Kibbo Kift propagated a form of green spiritualism, a pantheistic belief in a divine Logos that supports all creation. At the same time, this almost religious dedication was also underpinned by scientific findings, and the influence of Darwin's theories. Thus, followers of the movement were encouraged to study the natural world, including mineralogy, rock formation, and animal and herb lore. Birdwatching was promoted, as well as the protection of endangered species.[4] Additionally, camping sites were to be organized in a way that did not disrupt the natural world, as the lyrics of one Kibbo Kift song testify:

> Then here's to the Kin, with their camps so clean
> You can't find the spot where their tents have been;
> They know what they want and they say what they mean
> With their hoods and their staves in the morning.
> (*The Nomad*, September 1923, 43)

Hoods and staves referenced in the final line were also an important aspect of the Kindred's distinctive identity. Beyond their environmental conservation efforts, the Kibbo Kift emphasized a wanderer's way of life, the right to roam freely and the need to be self-sufficient. Hargrave carefully avoided the word "uniform" in descriptions of the Kibbo Kift costume; the characteristic hooded jerkin of his design harked back to a simpler era of Robin Hood and his merry throng. Kinsfolk were encouraged to make their own clothes, construct and paint their canvas tents, as well as carve their own totems (often representing animals) to express their individuality. *The Nomad*, another of their periodicals, offered advice on tent decoration, and on natural dyes to be obtained from "things of the Earth" (April 1925, 247). Shop-bought ready-made items were frowned upon: Hargrave's practical didactics involved learning by doing, and besides he believed that the creative impulse inevitably emerges from encounters with nature: "It was through this struggle with his environment that man at last evolved the cunning handicrafts and the power of expressing himself by art, music, writing and speech" (*Great War Brings It Home* 243).

---

4 Already in 1921, the group launched an official campaign to save the Yellow-Necked Woods Warbler and went on to protect many other bird species. Hargrave's account of his service at Gallipoli, *At Suvla Bay* (1916), testifies to his intense interest in the other inhabitants of the natural world, even in the direst of circumstances: "Queer creatures crept across the sands and into the old Turkish snipers' trenches; long black centipedes, sand-birds – very much resembling our martin, but with something of the canary in their colour. Horned beetles, baby tortoises, mice, and green-grey lizards all left their tiny footprints on the shore" (134).

Arts and crafts combined with hiking seemed to echo the utopianism of William Morris, and yet Hargrave insisted on the modern character of his movement. Annebella Pollen draws attention to this curious retro-futurist approach, highlighting Kibbo Kift's fruitful eclecticism which, it was hoped, would meet the challenges of the modern world:

> The intersection of past, present and future is of key significance to understanding Kibbo Kift's philosophy; to look back was not to reject the modern world but to revisit cultural history in order to develop the group into something they hoped would be distinctively avant-garde. (Pollen, "More Modern than the Moderns" 327)

Rather than break away from the past, Hargrave sought to weld some of the old ideas into a new synthesis. Likewise, in his artistic creations and designer practice, he searched for fresh forms of expression, which ultimately resulted in a striking, unmistakable visual identity of his group. The Kibbo Kift's aesthetic was characterized by futuristic geometric patterns and bold use of colour, while at the same it was gesturing towards the Robin Hood "outlaw" look, suggestive of freer, pre-nationalistic social structure. Celtic, Anglo-Saxon, Icelandic, and American Indian influences were welcome for a similar reason: their artwork was produced by peoples who did not know the nation-state but were loosely organized in tribes defined by communal bonds. They also had a kinship attitude to land and perceived themselves as interdependent with plants, animals, and the inanimate natural surroundings.

Shaped by nature-centric worldviews, the cultural practices of the Kibbo Kift were inevitably positioned on the margins of mainstream art currents. As Pollen points out, despite the richness of its arts, craft, design and performance, the Kinsfolk production was not, until recently, properly catalogued or displayed in exhibitions, nor retrospectively included in literary canons. Because it "circulated outside the privileged metropolitan elite," its status within modernism "appears tenuous" ("More Modern than the Moderns" 207). Nevertheless, Hargrave was personally acquainted with many visual practitioners (Augustus John, Maxwell Armfield, Jacob Epstein) and well familiar with the contemporary trends in art and design – Cubism, futurism, Bauhaus, Russian Constructivism, the costumes and scenography of the Ballets Russes. He popularized and democratized them, instructing Kinfolk how to decorate vestments for performances and ceremonies. Members of the tribe were even meant to act as living advertisements for their idealist mission on a daily basis, urged to wear colourful clothes, enhancing their vitality and proving they were unspoilt by the drabness of industrial society.

To the elite, upper-class modernists, Hargrave's project must have seemed as outlandish as the Kinfolk's striking vestments, and probably a bit too middlebrow. A list of Hargrave's fifty most dedicated adherents, provided by Cathy Ross and Oliver Bennett in the Museum of London album *Designing Utopia*, suggests that many members were art school teachers, architects, photographers, and performers. Some had academic interests, while others were authors of popular literature. Most pursued day jobs and therefore could be "dismissed as class deficient by those in more privileged positions" (Pollen, "More Modern than the Moderns" 334n). Nominally, the group had the endorsement of some high-profile artists and intellectuals: H. G. Wells, Havelock Ellis, Maurice Maeterlinck, Rabindranath Tagore, Julian Huxley, J. Arthur Thompson, and Patrick Geddes were on the Kibbo Kift advisory council, though that does not mean they were directly involved in the movement's activities. Yet Hargrave never really wished for his followers to blend in and be part of the current establishment. Likewise, superficial distinctions of class, race, and nation were not relevant to him. He believed that Kibbo Kift's techniques of personal development through woodcraft training conferred special status on its members: they were agents of rebirth, the new nobility working for the long-term goal of world unification. The select few were to become the hope of all humanity, for they would "blaze their own trail [...] the trail of nature – the trail of woods and fields and rivers" (Hargrave, *Great War Brings It Home* 310). Because the realization of their dream lay far off in the future, they were, in Hargrave's deepest conviction, "more modern than the moderns" (Hargrave qtd. in Pollen, "More Modern than the Moderns" 328).

Sometimes their channelling of an inner barbarian served as an inspiration for better-known avant-garde figures. As David Bradshaw has argued, D. H. Lawrence might have modelled the character of Mellors in *Lady Chatterley's Lover* on Hargrave, with his flamboyant, gaudy outfits and his seclusion in the woods, as well as economic beliefs that evoke Kibbo Kift's enthusiasm for social credit, a radical theory advocating for the redistribution of wealth. Lawrence was acquainted with Hargrave's ideas through Rolf Gardiner, a rural revivalist and an enthusiast of Morris dancing, who briefly joined the Kibbo Kift but later dismissed it as having "no real roots in the soil of England, no blood-contact with the living part of English earth" (Gardiner qtd. in Moore-Colyer 192).[5] The exchange of letters between Lawrence and Gardiner in the

---

5   Rolf Gardiner had an uneasy relationship with Hargrave. He joined the Kibbo Kift in 1924 but left the following year. In the 1930s, he developed Land Service Camps for Youth in northern Europe and forged numerous connections with German youth leaders. As Hargrave's and Gardiner's paths diverged, it became clear that the former celebrated nomadic rootlessness

years 1924–1928 suggests that both ultimately disagreed with Hargrave's idealistic dream of oneness and universal brotherhood among youth.[6] Indeed,
Hargrave's vision was transnational, closer to the utopianism of H. G. Wells
and his dream of world government and world culture. In a long list of points
to remember, *The Nomad* reminded the readers that

> The Kindred belong to one Country: The World.
> The Kindred have one loyalty – to mankind.
> A Kinsman trains himself physically – by Campcraft.
> A Kinsman trains himself mentally – to Think Internationally.
> (*The Nomad*, October 1923, 60)

Unlike peasants and farmers, who are tied to the land, Kinsfolk were cosmopolitan ramblers, freely exploring the countryside. In the early, idealistic phase
of their activity, questions of land ownership interested them only insofar they
could use somebody's grounds for camping. Invoking the medieval Domes-
day Book, Hargrave initiated a survey of the open spaces appropriate for the
practice of woodcraft. *The Mark* and *The Nomad* regularly published information about Kibbo Kift-friendly landowners in Britain and abroad. Plans were
made for the creation of Kin-Garth, a tract of land where the movement would
once set up "a central encampment and Althingstead" (*The Nomad*, September
1924, 195) for the training of world citizens. The group also established connections with like-minded formations across Europe: *The Republique Supra-
nationale* and *Union Fraternelle du Scoutisme* in France, The Czechoslovakian

---

of the wanderer, and the latter – the rootedness of the peasant. Gardiner was also doubtful
about the possibility of world unity and disliked Hargrave's fascination with H. G. Wells. For
a broader discussion of the Gardiner-Hargrave rift, see: Craven, J. F. C. *Redskins in Epping
Forest: John Hargrave, the Kibbo Kift and the Woodcraft Experience.* Doctoral thesis, University
of London, 1999, pp. 204–230.

6 In his letters to Gardiner, Lawrence grew increasingly critical of Hargrave; some scholars
argue that it was him who might have persuaded Gardiner to leave the Kibbo Kift (see Moore-
Colyer, 193). Having read Hargrave's publications, Lawrence claimed that he respected Hargrave, and that "[his] ideas are sound, but flesh and blood won't take 'em [...] he'll get no
further than holiday camping and mummery [...] But by wanting to rope in *all* mankind it
shows he wants to have his cake and eat it. Mankind is largely bad, just now especially – and
one must hate the bad, and try to keep what bit of warmth one can, among the few decent"
(Lawrence 1034). He also dismisses Hargrave's "dummy teat of commiseration," criticizing
"all this blasted snivel of hopelessness and self-pity and 'stars' and 'Wind among the trees'
and 'camp fires' and witanagemotery – It's courage we want, fresh air and not suffused sentiments'" (Lawrence 800).

Woodcraft League, the short-lived Union of Free Scouting (*Zjednoczenie Wolnego Harcerstwa*) in Poland, various *Bünde* in Germany, scouts from Holland, Belgium and Italy. Small branches of the Kibbo Kift were set up in Russia, Australia and Kenya. The Kibbo Kift Declaration was rendered into several languages, including Esperanto and Ido, and Hargrave's books reached a wider audience in translated and adapted versions.

Unfortunately, as the 1920s progressed, the Kibbo Kift's expansion stalled. Emphasis on country pursuits and international collaboration suddenly lost relevance and even became suspect. Hargrave's foreign contacts drew the attention of MI5, which feared possible infiltration by dangerous ideologies. The Kinfolk's map-making practices were similarly scrutinized, as they might be interpreted as espionage activities. Within the movement itself, there was increasing pressure for more decisive political commitment. In 1924, this internal tension culminated in a significant schism known as the "Rift in the Kift." The left-wing grouping, led by Leslie Paul, challenged Hargrave's authority and broke away to form the Woodcraft Folk – an organization that still exists today,[7] affiliated with the Co-operative movement.

The Rift not only fragmented the Kibbo Kift but also reflected broader societal shifts. The optimism and idealism of the early 1920s were giving way to a more pragmatic and politically charged atmosphere of the Depression years. Participation in the Kindred activities was, after all, a time-consuming form of leisure, and in the period of crisis fewer people could embrace this type of lifestyle. Hargrave developed an interest in the theories of Major C. H. Douglas and became convinced that the Kibbo Kift's new focus should be building a civilization based on complete economic security for the individual. Preoccupation with social credit gradually replaced woodcraft and personal development: Hargrave's organization mutated into The Green Shirts Movement, and later a political force with a decisively militant edge, The Social Credit Party of Great Britain and Northern Ireland. While camping remained a component of the organization's activities, the emphasis shifted to training shock-troopers, ready to confront Oswald Mosley's Blackshirts and communist Redshirts in the numerous 1930s riots in London and other British cities. In the words of Andrew Marr, "the rambling stopped and the marching began," as Hargrave's Green Shirts became "the most

---

7   In 2025, Woodcraft Folk celebrates its centennial. The movement's ongoing commitment to cooperation, peace, and environmentalism echoes some of the Kibbo Kift's original aspirations. The Woodcraft Folk promotional film, prepared specially for the occasion, explores some of this legacy: https://woodcraft.org.uk/about-woodcraft-folk/our-history/.

disciplined marchers of all, heckling at meetings, throwing bricks through the chancellor's window, tramping through the streets under their distinctive flags, running demonstrations and 'street patrols'" (288–289). Like the modernist subjective turn, the movement's earlier focus on spiritual searching and health through nature seemed increasingly out of step as the world approached another global war. Hargrave and his followers had to look outward, forsake individualism, dress and act uniformly, in order to swell their ranks and gain greater visibility in politics. This initiative, initially successful, lost momentum after the 1936 Public Order Act, which prohibited the wearing of political uniforms in public. The movement further dwindled with the outbreak of World War II, as national priorities shifted once again, and many members were drafted into the military.

In *The Confession of the Kibbo Kift*, Hargrave reflected retrospectively upon the various youth and nature movements that emerged in the early 1920s. With a tinge of regret, he noted that their enthusiasm was insufficient to create lasting change:

> It is not unfair to say that these movements (for there were many), which flourished during the years 1920–1924, were an escape, tinged with a vague romanticism. There was something of the troubadour tradition – strolling with a guitar and a rucksack – and one begins to wonder whether there was anything upon a purely logical plane to balance this emotional uprising.
>
> By the year 1925 it was possible to discern a waning of the original aureole of effulgent idealism [...]. True, they went back, as they thought, to fundamental impulses – that is, they made a conscious attempt to merge the fractured individual ego into the unconscious, un-individualised Nature throb – but they did not go back far enough. (Hargrave, *Confession* 38)

Despite the fact that the Kibbo Kift Kindred never had as large a following as Baden-Powell's Boy Scouts internationally or the Woodcraft Folk in Britain, Hargrave's vision still carries the potential to resonate with twenty-first-century oppositional art and culture. He left a blueprint for a better society, offering a unique perspective on the imaginative use of non-working time. A hundred years later, in the uncertainties of contemporary politics, still affected by the disarray of the 2008 recession, the fallout from the COVID-19 pandemic, and the war in Ukraine, we hear echoes of Hargrave's ideas among the protest activism in the style of the Occupy movement, in the calls for sustainable living and mental health awareness, or in the trial schemes implementing universal basic income. Hargrave's principles of personal development, communal living, and a deep connection with nature appear particularly pertinent in modern times, and his legacy underscores the enduring relevance of grassroots movements in enacting social transformation.

Bibliography

Bradshaw, David. "Red Trousers: *Lady Chatterley's Lover* and John Hargrave." *Essays in Criticism*, vol. 55, no. 4, 2005, pp. 352–373.

Claeys, Gregory. *Utopianism for a Dying Planet: Life After Consumerism.* Princeton University Press, 2022.

Craven, J. F. C. *Redskins in Epping Forest: John Hargrave, the Kibbo Kift and the Woodcraft Experience.* Doctoral thesis, University of London, 1999.

Drakeford, Mark. *Social Movements and Their Supporters: The Green Shirts in England.* Macmillan, 1997.

Hargrave, John. *At Suvla Bay. Being the Notes and Sketches of Scenes, Characters and Adventures of the Dardanelles Campaign.* Constable & Company Ltd., 1916.

Hargrave, John. *The Great War Brings It Home: The Natural Reconstruction of an Unnatural Existence.* Constable & Company Ltd., 1919.

Hargrave, John. *Harbottle: A Modern Pilgrim's Progress from This World to That Which Is to Come.* Duckworth, 1924.

Hargrave, John. *The Confession of the Kibbo Kift: A Declaration and General Exposition of the Work of the Kindred.* Duckworth, 1927.

Huddersfield Daily Examiner, 14 March, 1921. British Newspaper Archive.

Lawrence, D. H. *The Collected Letters of D. H. Lawrence,* vol. II, edited by Harry T. Moore, Heinemann, 1970.

Marr, Andrew. *The Making of Modern Britain.* Macmillan, 2009.

McCarthy, Jeffrey Mathes. *Green Modernism: Nature and the English Novel, 1900–1930.* Palgrave Macmillan, 2015.

Moore-Colyer, R. J. "Rolf Gardiner, English Patriot and the Council for the Church and Countryside." *The Agricultural History Review,* vol. 49, no. 2, 2001, pp. 187–209.

Pollen, Annebella. *The Kindred of the Kibbo Kift: Intellectual Barbarians.* Donlon Books, 2015.

Pollen, Annebella. "More Modern than the Moderns: Performing Cultural Evolution in the Kibbo Kift Kindred." *Being Modern: The Cultural Impact of Science in the Early Twentieth Century,* edited by Robert Bud, Paul Greenhalgh, Frank James and Morag Shiach, UCL Press, 2018, pp. 311–336.

Ross, Cathy, and Oliver Bennett. *Designing Utopia: John Hargrave and the Kibbo Kift.* Philip Wilson Publishers, 2015.

The Mark. An Independent Kibbo Kift Magazine, June 1922–May 1923. LSE Archives.

The Nomad, June 1923–May 1925. LSE Archives.

"Woodcraft Folk Heritage Film." Woodcraft Folk Website. https://woodcraft.org.uk /about-woodcraft-folk/our-history. Accessed 28 February, 2025.

CHAPTER 5

Beyond Dualism: Human–Nature Connections as Anthropomorphic Interaction in Contemporary Environmental Literature

Tereza Dědinová

Abstract

The fundamental discourse of the Anthropocene includes critical reflections on the culture–nature relationship and the possibility of overcoming this widespread dualism. The chapter focuses on the challenges to the culture–nature divide in contemporary environmentally-oriented literature, represented by the works of Paolo Bacigalupi, Richard Powers and Terry Pratchett. The analysis emphasises the subversion of dualistic representations of the culture–nature relationship in the selected works while arguing that the subversion is achieved specifically through the anthropomorphism of nature as a disanthropocentric strategy and the representation of the mesh as "the interconnectedness of all living and non-living things" (Morton 2010). As Marco Caracciolo claims in *Narrating the Mesh* (2021), anthropomorphism may be a successful way of opposing dualism and human exceptionalism. This chapter draws especially on Gabriella Airenti's (2019) distinction between anthropomorphic belief and anthropomorphic interaction, where the latter is understood as an essential strategy for relating to the nonhuman world. The theoretical insights underlie the subsequent discussion of three literary texts. Whereas Bacigalupi's science fiction novella "The People of Sand and Slag" (2004) illustrates the harmful consequences of anthropocentrism and lack of anthropomorphic interaction, Powers's *The Overstory* (2018) and Pratchett's Tiffany Aching series (2003–2015) conjure up imaginary worlds in which anthropomorphic interaction fosters communication and connection between the human and the nonhuman. By doing so, these works partially overcome the nature-culture dualism and evoke a sense of consoling, uplifting interaction.

Keywords

anthropomorphic belief – anthropomorphic interaction – nature–culture dualism – the mesh – environmental crisis

© KONINKLIJKE BRILL BV, LEIDEN, 2026 | DOI:10.1163/9789004744110_007

1   Introduction

On 11 November 2021, the eminent British naturalist and documentary film-maker David Attenborough, speaking at the UN Climate Change Conference in Glasgow, asked: "Is this how our story is due to end? A tale of the smartest species doomed by that all too human characteristic of failing to see the bigger picture in pursuit of short-term goals" (00:03:16–00:03:32). It is precisely this bigger picture, in other words, grasping global phenomena in their complexity and interconnectedness, which includes the relationship between nature and culture, that poses a fundamental problem in understanding the current environmental crisis of which climate change is a part.

This chapter analyses strategies for overcoming the culture–nature divide in environmentally-oriented literature, which is understood here broadly as writing which comments on the mutual connections between nature and humanity. In particular, the chapter focuses on the anthropomorphism of nature as a disanthropocentric strategy and mesh-conscious representation. First, I invoke the matter of dualism and its critique from the ecocritical and cognitive perspectives. Then, I focus on the central concepts of the mesh, as introduced by Timothy Morton, and anthropomorphism from a cognitive perspective. Before employing anthropomorphism in readings of three literary works chosen for their specific representations of the relationship between humans and the nonhuman world, anthropomorphism is further discussed in terms of its potential disanthropocentric effects and the distinction between anthropomorphic belief and interaction proposed by Gabriella Airenti (2019).

2   Challenging Dualism

The environmental humanities and deconstruction expert Timothy Clark (2015) writes that in the case of climate change "there is no simple or unitary object directly to confront, or delimit, let alone to 'fix' or to 'tackle'. There is no 'it', only a kind of dissolution into innumerable issues" (10). Even when reducing the matter of environmental crisis solely to global warming, every individual might consider some of its aspects more or less severe depending on her experience, history, location and knowledge. While some effects of global warming literally stick to our skin, others may seem to be a future problem or a problem pertaining to distant lands – to tackle it and the Anthropocene in general, humanity, according to Morton, needs to develop species-consciousness, in other words, a capability to "consider its impact as a totality upon the whole planet" (16).

According to contemporary environmental scholars, acknowledging the broader context of ongoing processes is significantly complicated by the persistent concept of the barrier between humans and the environment, culture and nature, subject and object.[1] However empirically indefensible, the supposed culture–nature divide has strong inertia, particularly in Western philosophy, and in its consequences it directly impacts the state of the planet (Arias-Maldonado 33–35). In *Environment and Society: Socionatural Relations in the Anthropocene* (2015), Arias-Maldonado rejects Descartes's notion of ontological culture–nature dualism and understands it as an emerging, historical phenomenon originating in socionatural relations: "the human–nature division has become real through processes such as the separation between urban and rural life, the exclusion of nature as an agent of human socialization, or the increasing digitization of social processes" (34). Building on Val Plumwood's ecofeminist analysis, Greg Garrard points out that the mere differentiation of humans from nature is not the heart of the problem. That is created by "*alienated* differentiation and denied dependency: in the dominant Euro-American culture, humans are not only *distinguished* from nature, but *opposed* to it in ways that make humans radically alienated from and superior to it" (25, emphasis in the original). Moreover, from the mechanistic materialism advocated, for example, by Thomas Hobbes, René Descartes, and Pierre Simon Laplace, our culture inherited an understanding of nature as "inert matter acted upon by stable physical laws" (Arias-Maldonado 24). The resulting dichotomy is thus hierarchical, putting humans above nature (Arias-Maldonado 24). When summarising arguments against dualism, Marco Caracciolo (2021) refers to Jane Bennett's *Vibrant Matter* (2010), a crucial book for material ecocriticism, which claims that there is nothing passive or inert in the natural world. Bennett portrayed the human and nonhuman relationships as an "intricate dance with each other" (31), an expression rendering both the simultaneity and agency of the coupling parties. This strongly antidualistic approach connects the material ecocriticism with the second generation of cognitive sciences,[2] which maintains that:

> a living organism and its environment arise simultaneously, in a structural coupling in which the features of the environment are defined by the

---

1   E.g. Alaimo 2010, Caracciolo 2021, Clark 2014, Garrard 2004, Heise 2008, Morton 2010, and others.

2   The second-generation approaches emphasise the crucial role of the embodiment of mental processes and their extension into the world through socio-cultural practices and material artifacts. For details, see Kukkonen and Caracciolo 2014.

organism's individual and collective (evolutionary) history of interactions with the world: in this sense, the environment does not preexist the organism but is cocreated by it. (Caracciolo, *Narrating the Mesh* 120)

As Caracciolo argues, the human/nonhuman dualism imbricates with the subject/object dualism since "our understanding of subjectivity is based on human subjectivity, with animals occupying an uneasy, and often contested, position on the fringes of the subject" (*Narrating the Mesh* 120). Challenging the rut of conventional thinking and refusing the notion of the nonhuman world as "insentient matter" (*Narrating the Mesh* 120), material ecocriticism seeks "more potent, more complex understandings of materiality" (Alaimo 2), bringing humans, their material creations and ideas, and nature closer together[3] and highlighting the role of story in understanding the world.

The inertia of dualism seems paradoxical, particularly in recent decades. It has become clear that nature, in the sense of a wilderness unaffected by humans, no longer exists on our planet. The famous photograph of the plastic bag at the bottom of the Mariana Trench (Morelle) or the presence of microplastics in the atmosphere can serve as telling proof. As Timothy Clark (2014) sums up: "To be standing in the remotest desert or ocean is now, in a sense, still to be breathing the enclosed air of a vast human crowd" (79). To characterise this reality, the traditional notion of nature seems increasingly problematic, and Timothy Morton in *Ecology Without Nature* (2007) directly suggests forgetting it. Clark (2011) argues that today different environments, some less affected by humans than others, have taken the place of nature (*Cambridge Introduction to Literature and the Environment* 6). Even if we stick with the label "nature," the legitimacy of the allusions we have long associated with it and which still dominate its cultural representations is lost. It is no longer (or perhaps never was) a haven to escape to from the confusion and anxieties of civilisation, a place of initiation and adventure, a framing space whose cycles build on each other in predictable and reliable ways.

As an alternative way of describing the relationship between nature and humans in the Anthropocene, new terms have emerged that denote the relationality, unpredictability, and inherent mutability of hybrid forms of reality involving the environment, humans and their products, material objects, and

---

3   In their influential *Material Ecocriticism* (2014), Serenella Iovino and Serpil Oppermann adopt the expression "naturecultures" proposed by Donna Haraway in *When Species Meet* (2007) to confront the divide between the animal and the human.

ideas.[4] Morton (2010) proposes the term *mesh*, which I will employ in this chapter; Stacy Alaimo (2010) *transcorporeality*, and Manuel DeLanda (2016) works with the term *assemblage*.

In the last decades, the number of ecocritical studies focusing on literary strategies for grasping the complex reality of the world and its experience in the Anthropocene, especially in the context of the environmental crisis, has grown significantly. Alaimo (2010), Caracciolo (2020; 2021), Clark (2014; 2019), Heise (2008), Iovino and Oppermann (2014), Weik von Mossner (2020) – to name just a few – unveil the mosaic of diverse approaches through which literature transcends dualism and mediates the intricate reality of the relationship between humans and the world. In this chapter, I intend to build on their findings when focusing on artistic strategies of hybridisation of humanity and nature through anthropomorphism of nature as a disanthropocentric and mesh-consciousness-building strategy.

3   Anthropomorphism of Nature and Mesh-Consciousness

According to classical definitions, anthropomorphism attributes human features, characteristics or form to nonhuman facts and phenomena. Anthropomorphism of nature and, in particular, animals is often seen as problematic (Grasso et al. 2020; Ganea et al. 2014),[5] and seems to occupy an inherently anthropocentric stance, subordinating nature to human perspectives and concepts and renouncing any attempt to reflect on the nonhuman from other than the human viewpoint. While "anthropomorphism and anthropocentrism have long worked in tandem" (Moore 12),[6] recent empirical research indicates

---

4 In Clark's words: "incalculable connection between bodies, human and nonhuman, across and within the biosphere (food, water, nutrients but also toxins and viruses), with a sense of both holism and, increasingly, entrapment" ("Nature, Post Nature" 80–81).
5 According to Ganea et al. (2014) anthropomorphism contributes to the dissemination of mistaken ideas about biological processes, which, notably in children, can lead to inappropriate behaviour towards animals. In general, as Grasso et al. (2020) claim, the nonrealistic – anthropomorphised – representation of animals in mass media can have considerable harmful effects. The authors invoke Litchfield's (2013) research into the demonisation of certain species in mass media with its potentially negative consequences for their conservation and conclude that "anthropomorphism seems to play a critical role in affecting public perception towards wildlife" (Grasso et al., 17).
6 In his monograph *Ecology and Literature: Ecocentric Personification from Antiquity to the Twenty-first Century* (2008), Bryan L. Moore explores anthropomorphism in literature, tracing its use from antiquity. He challenges the perception that anthropomorphism is outdated and anthropocentric, emphasising its ecocentric potential.

quite a different perspective, showing that anthropomorphism fosters pro-environmental behaviour. Likening nature to humans and their experience, as in Al Gore's Nobel Lecture – "As a result, the Earth has a fever" (2007, qtd. in Tam, Lee and Chao 514) – can influence significantly how "people relate to and behave toward nature" (Tam, Lee and Chao 520). More specifically, Tam, in a study of environmental guilt, concludes:

> Individuals who anthropomorphize nature are more likely than those who do not to experience guilt when considering environmental degradation, and this guilt feeling, in turn, is associated with more engagement in pro-environmental behavior. (Tam, "Anthropomorphism of Nature" 16)

In the field of literary studies, Serenella Iovino and Serpil Opperman adopt a similar attitude when claiming that anthropomorphism "can even act against dualistic ontologies and be a 'dis-anthropocentric' stratagem meant to reveal the similarities and symmetries existing between humans and nonhumans" (8). Similarly, Marco Caracciolo (*Narrating the Mesh* 160) builds on Jane Bennett's well-known claim that anthropomorphism, cultivated as a perception of the echoes of human agency in nonhuman nature, is the antithesis of "the narcissism of people in charge of the world" (Bennett 2010, XVII). Caracciolo analyses the conditions under which anthropomorphism goes beyond mere appropriation and demonstrates that anthropomorphism successfully challenges dualism and human exceptionalism, provided it emphasises the bidirectionality and nonlinearity of analogies between the human and the nonhuman. Drawing on his exploration of conceptual metaphors, Caracciolo identifies functional anthropomorphism as one that brings new insight into both source and target domains, and avoids "projecting the human in a way that erases the distinctiveness of the nonhuman as well as the nonlinearity of their entanglement" (*Narrating the Mesh* 176). For instance, in the conceptual metaphor mentioned by Al Gore, "As a result, the Earth has a fever," the Earth is a target domain, while fever is a source domain referencing living organisms. Most recipients are likely to connect the notion of fever with the manifestation of illness in humans and the need for treatment. At the same time, the implication of the metaphor may be an approximation from the other side, from the target to the source domain – a shift in the perception of living organisms inhabiting the feverish Earth. When focusing on humans, we are reminded of various forms of causality (the consequences of human actions as a disease that manifests as an increase in the Earth's temperature) and the possibility of relieving the Earth and, therefore, ourselves – especially in the context of Al Gore's speech.

The concept of anthropomorphism as a means of challenging the dualistic ontology and human exceptionalism also supports the research collected in *The Cognitive Underpinnings of Anthropomorphism* (2019). In its opening study, Gabriella Airenti distinguishes between, on the one hand, anthropomorphic belief, defined as the attribution of human characteristics to nonhuman entities and events and, on the other hand, anthropomorphic interaction independent of the beliefs about the nonhuman that people might entertain. According to Airenti, anthropomorphism is an essential strategy for relating to the world. When people anthropomorphise natural phenomena, such as illnesses, in effect, it is "not the event itself that is anthropomorphized but rather the relation that a person establishes with it" (9). While doing so, they employ the exact cognitive mechanisms as when interacting with other people (16). In other words: "Fundamentally, anthropomorphism is a way of relating with a non-human entity by addressing it as it were a human partner in a communicative situation" (14). Arguing against the Piagetian view of animism (which includes anthropomorphism in Piaget's definition) and referring to a number of empirical studies, Airenti posits anthropomorphic interaction as "the basis of any form of anthropomorphism" (16) and rejects its interpretation as a childhood attitude, a form of irrational reasoning from which a child retreats when developing causal thinking (13).

In her study of anthropomorphism in human-animal interactions, Véronique Servais takes a similar approach, leaning on the definition of anthropomorphism as "relating to other animals as subjects and agents, with feelings, intentions, needs, and so on" (38) and – like Kay Milton – highlights its interactional nature: "personhood is not a property of something, it emerges out of what something does in relation to others" (41). To sum up, from this perspective anthropomorphism is the opposite of the objectification of the nonhuman based on the notion of human exceptionalism. To anthropomorphise means to open yourself to interaction, negotiation and relationship. Such a communicative situation does not encourage simple isomorphic projection. Through likening each to each other, both human and nonhuman parties move from their initial positions, bridging the culture–nature divide and transcending themselves into the mesh. Anthropomorphic interaction thus has a potential to contribute to the building of mesh-consciousness, i.e. to the awareness of "the interconnectedness of all living and non-living things" (Morton, *Ecological Thought* 28).

From a literary perspective, we can observe the blurring of the imaginary camera's viewfinder away from humans towards nonhuman objects and phenomena, including the environment. The human is not automatically at the centre of the story as the only significant actor or focaliser; instead, it is depicted as part of nature subjected to its rules, not elevated above it. Neither human agency nor the human timescale dominate the narrative, which challenges the bias toward human and human-like characters. This strategy rejects the convention of putting humans in the role of actants around which the plot revolves, and nonhumans as objects – "tools to further human ends, or a backdrop to human-centered events" (Caracciolo, *Narrating the Mesh* 99). In other words, paradoxically, anthropomorphic interaction supports "letting go of human exceptionalism and fostering a sense of our vital interdependency with the nonhuman world" (*Narrating the Mesh* 3).

4   No Nature, Only Culture

How does environmentally-oriented literature represent the human–nature relationship through the anthropomorphism of nature and how does it foster mesh-consciousness? Let us begin with a zero variant, a fictional world where people have severed as many of their ties to nature as possible and thrive on a devastated Earth. The science fiction novella "The People of Sand and Slag" by renowned climate fiction writer Paolo Bacigalupi was first published in 2004, nominated for the Hugo Award and the Locus Poll in 2005 and the Nebula Award in 2006. Nonhuman life is practically extinct in an unspecified far future, with the rare exceptions of animals in specialised zoos and laboratories. No non-genetically modified organism can survive long on a scorched, polluted, and human waste-covered land. Humanity, on the other hand, has adapted to the new conditions. Due to the mastery of advanced bioengineering technologies, human bodies are practically indestructible: they heal quickly from even the most severe injuries, grow body parts, and draw nutrients from odd sources, sand and slag included. However, humans are not entirely indifferent to their environment and, in the story, enjoy vacations at the oceanside. They still draw consolation from being outdoors, from interacting with the remnants of nature represented by the polluted ocean and trash-strewn beach. Passages depicting the three central characters' stay in Hawaii ingeniously combine holiday clichés with a chilling image of a trashed paradise:

> It was good to get out of the northern cold and into the gentle Pacific. Good to stand on the beach, and look out to a limitless horizon. Good to walk along the beach holding hands while black waves crashed on the sand. Lisa was a good swimmer. She flashed through the ocean's metallic sheen like an eel out of history and when she surfaced, her naked body glistened with hundreds of iridescent petroleum jewels. When the Sun

started to set, Jaak lit the ocean on fire with his 101. We all sat and watched as the Sun's great red ball sank through veils of smoke, its light shading deeper crimson with every minute. (Bacigalupi)

The encounter between humans and nature is represented by the central event of the novel when the group of friends find a malnourished, injured and exhausted dog – a genuine curiosity from their point of view, and a reminder of the evolution that humanity has undergone: "It's hard to believe we ever lived long enough to evolve out of that. If you chop off its legs, they won't regrow" (Bacigalupi). The summoned biologist explains to them the interconnected-ness of all life, its interdependence and fragility, something the new humans have broken and forgotten. The way back would be extremely complicated, even if one wanted it: "Re-creating the web of life isn't easy. Far more simple to release oneself from it completely than to attempt to re-create it" (Bacigalupi). All nature has become impossibly brittle compared to the indestructible and self-restoring humanity. Nature is fragile, whether it is the living body of a dog, or a stone, which we associate more with hardness and resilience: "It's as deli-cate as rock. You break it, and it never comes back together" (Bacigalupi). The characters of the story care for the animal briefly and feel the reverberations of something unknown: an unnamed emotion, the warmth and consolation of a relationship with a nonhuman being, but they do not overcome their self-ishness and kill the dog rather than accepting responsibility for its survival. Altogether, the people in the story seem to be a parody of perfect adaptation, losing the very ability to care and empathise together with their vulnerability and dependence on the web of life. The novella is a telling example of a broken relationship between humans and nature – people have ripped the mesh, and they have placed themselves at the centre of the world. Although meeting the dog elicits new feelings and disturbing questions in Chen, the narrator and most sensitive group member, this hint of an anthropomorphic interaction with the dog is too weak to make a difference. It only leaves an uncomfortable uncertainty: "If someone came from the past, to meet us here and now, what do you think they'd say about us? Would they even call us human?" (Bacigalupi)

"The People of Sand and Slag," depicting an extreme separation of people from nature, offers a powerful warning against solely human-centred solutions to environmental crises. For the heavily modified people capable of surviving and even thriving on devastated Earth, the memory of the web of life – and the mesh as a whole – is merely a curiosity; they see interdependency as a disad-vantage, and animals as an evolutionary dead end.

Ironically, the mesh still exists in some respects in their world but is dra-matically altered and depleted of nonhuman agents. Humans are still part of the world and depend on it to meet their basic needs, and they also take refuge in the poor remnants of nature to renew their strength. However, they do not acknowledge the rejuvenating and consoling potential of nature in its complexity and fail to understand the need for reciprocity and consideration for the nonhuman world. Alone in their physical indestructibility, humans regard themselves as the pinnacle of evolution, unaware of all they have deprived themselves of. The chance of anthropomorphic interaction is not acted upon in the story, so it does not change anything. While the characters have no real relationship to nature, nor do they know what life was like in a fully mesh-connected world, the reader is implicitly invited to compare the dreary fictional world with the beauty of nature in her own world, as well as to reflect on its endangerment. The novella's effect on the reader is further enhanced by Bacigalupi's account of the characters' encounter with the dog, which for many no doubt personifies selfless love and loyalty, as a representative of dying and betrayed nature. Pure anthropomorphic belief, i.e., an isomorphic projection that does not open up the possibility of communication, is depicted ironically in Bacigalupi's novella, whose characters have difficulty understanding that a genetically unadapted dog cannot heal as quickly as they do. Their ignorance and anthropocentric attitude are conditioned by their lack of familiarity with other life forms.

The novella portrays a bleak picture of an anthropocentric approach in which anthropomorphic belief is manifested, but anthropomorphic interaction is dramatically weakened. While the humans in the story occasionally feel a connection to the natural world around them – the ocean, the beach, and the dog – they ultimately exploit these resources for their pleasure without regard for the needs of the nonhuman. As a result, the possibility of finding renewal, redemption, consolation and regeneration of their humanity through the connection to nature remains unfulfilled.

5   Of Trees and Man

The Overstory (2018) by Richard Powers attracted considerable critical attention: it was shortlisted for the Man Booker Prize in 2018 and awarded the Pulitzer Prize for Fiction in 2019 and the William Dean Howells Medal in 2020. The novel's title is a pun, referring to the intertwined treetops and to the central narrative that covers the individual stories of the nine human characters and many trees. This connection is also reflected in the book's structure, whose sections are named after parts of a tree: roots, trunk, crown, and seeds. Powers interweaves the stories of trees and people on many levels, points out their interdependence and contrasts the shortness of human life with the longevity of trees. Humans and trees are represented as thoroughly enmeshed. People and trees occupy the same narrative space, living next to each other, interacting, and communicating. Even when the human characters are unaware of all the relations, readers are made to perpetually maintain a bigger picture:

> He wonders: What makes the bark twist and swirl so, in a tree so straight and wide? Could it be the spinning of the Earth? Is it trying to get the attention of men? Seven hundred years before, a chestnut in Sicily two hundred feet around sheltered a Spanish queen and her hundred mounted knights from a raging storm. That tree will outlive, by a hundred years and more, the man who has never heard of it. (Chap. "Roots: Nicholas Hoel")

Most of the human characters live in close interaction with trees: the Hoel family, whose founder plants a chestnut tree in Iowa and for generations watches the growth and eventual death of a majestic tree (the genus was ravaged in America by chestnut blight in the early 1900s), or a daughter of an immigrant who plants a mulberry tree in the backyard of his new home after fleeing China. One of the central characters, Patricia Westerford, is based on the life and work of forest ecologist Suzanne Simard, famous for her research on communication and interactions between trees and other plants in the ecosystem through the underground network created by roots and fungi. Being aware of the hidden life of trees from childhood – "plants are willful and crafty and after something, just like people" (Chap. "Roots: Patricia Westerford") – adult Patricia writes *The Secret Forest*, a book conspicuously echoing the real-world works by Colin Tudge and Peter Wohlleben.[7] The opening passages of *The Secret Forest*, the book within the book, bring together the central theme both of the fictive book and *The Overstory*, that is the closeness of humans and trees: "You and the tree in your backyard come from a common ancestor. A billion and a half years ago, the two of you parted ways. But even now, after an immense journey in separate directions, that tree and you still share a quarter of your genes" (Chap. "Roots: Patricia Westerford"). Patricia conveys her striking insights into the life of trees and significantly disrupts fixed ideas of trees as relatively simple, more or less isolated organisms that cannot

---

7   Colin Tudge, *The Secret Life of Trees: How They Live and Why They Matter* (2005); Peter Wohlleben, *The Hidden Life of Trees: What They Feel, How They Communicate* (2016).

communicate with humans. Her words in *The Secret Forest* "made [trees] real" (Chap. "Crown") to its readers, transgressing what her father called Adam's curse, the fact that people are "plant-blind." *The Overstory* often addresses the human blindness to nature, challenging the limits of human-centred perception. As Patricia writes in her book:

> No one sees trees. We see fruit, we see nuts, we see wood, we see shade. We see ornaments or pretty fall foliage. Obstacles blocking the road or wrecking the ski slope. Dark, threatening places that must be cleared. We see branches about to crush our roof. We see a cash crop. But trees – trees are invisible. (Chap. "Crown")

Powers's novel strives to make trees visible, to overcome human biases and provide its readers with the bigger picture of the culture–nature relationship in a world where people are not at the centre of everything. Multiple strategies support this, one of them being a representation of trees as actants with the use of active verbs – e.g. "Tree saved your life" (Chap. "Roots: Douglas Pavlicek") – thus, as living beings capable of reciprocity and agency; they have their needs, goals, and strategies of communication. Reciprocity and communication between humans and trees often straddle the edge of anthropomorphic interaction and belief, as characterised by Airenti. The human characters feel an intimacy with the trees that translates into a need to defend them against human exploitation. In passages reflecting the active resistance of environmentalists against the logging of old-growth redwood forests, people sacrifice not only comfort and money; they often risk and, in some cases, lose their own lives. Individual trees are anthropomorphised, portrayed as persons with names and history, protected and mourned by their human companions. The anthropomorphism of trees enables the establishment of a relationship between humans and trees, but it does not reduce the tree to a human-like entity. On the contrary, together with the wealth of information about the life of trees, it inspires readers to attempt to become aware of trees in their intricacy and uniqueness. Also, people liken themselves to trees; tree-rights activists choose new names based on the different types of trees to affirm their new mesh-conscious selves symbolically. However, not only do people name individual trees, but they also break the imaginary barrier between nature and culture when they try to understand trees in their complexity, and they succeed only through cultural concepts (which are themselves a reflection of the relationship with nature). For example, when activists Nick and Olivia come to Mimas, a giant sequoia in whose canopy they will spend a year to save it (unsuccessfully) from being cut down by loggers, Nick responds to the encounter as follows:

> The tree runs straight up like a chimney butte and neglects to stop. From underneath, it could be Yggdrasil, the World Tree, with its roots in the underworld and crown in the world above. [...] Two more trunks flare out higher up the main shaft. The whole ensemble looks like some exercise in cladistics, the Evolutionary Tree of Life – one great idea splintering into whole new family branches, high up in the run of long time. (Chap. "Trunk")

Olivia's reaction is a reminder of the dependence of people on trees when her gesture of spreading her arms against the bark of the tree is described as "she's like a flea trying to hug its dog" (Chap. "Trunk"), and at the same time of the unevenness of the people-trees relationship in the human-dominated world: "I can't believe there's no other way to protect this thing except with our bodies" (Chap. "Trunk"). Olivia is also one of the human protagonists in *The Overstory* who hears trees talking to her directly. While the voices in her head can be interpreted as after-effects of the nearly fatal accident that opened her eyes to nonhuman life, other passages where humans, when distancing themselves from the hustle and bustle of civilisation, discover the ability to listen and understand the trees, to learn "the foreign language" (Chap. "Roots: Patricia Westerford"), validate Olivia's experience as genuine human–nature communication.

Anthropomorphism as a disanthropocentric means strongly manifests in *The Overstory*. Powers depicts the complex interdependence of people and trees on many levels, from the purely practical to the symbolic and spiritual. Trees in the novel function as part of the ecological mesh and as sources of consolation, offering emotional and spiritual regeneration to human characters. Through this, Powers connects environmental literature with more profound existential questions, portraying nature not as a passive backdrop but as an active partner in human life. Although the other elements of the mesh, save for the people and trees, are sidelined in the novel, *The Overstory* is a powerful depiction of the intertwining of nature and culture and a strong statement against dualism. The dominant anthropomorphic interaction approaches anthropomorphic belief when characters communicate with trees directly. The condition for both is openness – people moving away from civilisation open themselves up to sharing with nonhumans, appreciating their beauty and essence. Non-trivial knowledge of the lives of trees, their needs and agency mediated throughout the story plays a significant role. Knowledge and emotions, the ability to empathise with other-than-human nature, are the platform for building a relationship between culture and nature.

People gain far more from nature than they give back, and many of Powers's characters realise this. The novel's fictional world is dominated by a society disconnected from nature, and the characters striving to protect the trees, able to experience the interconnectedness of humans and trees, emerge as strange individuals misunderstood by others. Despite all this, the novel can be considered hopeful because it mediates possible interspecies communication and anthropomorphic interaction. The people in the novel derive comfort, consolation and a sense of meaning from their interactions with the trees. As the novel's opening chapter, in which a pine tree speaks to a woman leaning against its trunk, demonstrates, humans perceive woefully little:

> That's the trouble with people, their root problem. Life runs alongside them, unseen. Right here, right next. Creating the soil. Cycling water. Trading in nutrients. Making weather. Building atmosphere. Feeding and curing and sheltering more kinds of creatures than people know to count. A chorus of living wood sings to the woman: If your mind were only a slightly greener thing, we'd drown you in meaning. (Chap. "Roots")

While overcoming deeply ingrained anthropocentric perspectives is a multifaceted challenge, fostering an awareness of the mesh – understood as the intricate, dynamic interconnectedness of all living and non-living entities – can open pathways to more profound and reciprocal relationships with the natural world. Such relationships may offer humans new ways of understanding their place in the world, though this process often requires navigating tensions and rethinking long-standing assumptions.

6   Hills in Her Bones, Soul of the Land in Her Head

To call Terry Pratchett's Tiffany Aching series (2003–2015)[8] environmentally-oriented literature might seem a stretch. However, the five novels that follow the growing up of the central character of a young witch in the rural landscape of Chalk, set in a fantasy world alluding to the chalk escarpment of the Chiltern Hills in England, focus intensely on the reciprocal relationship between Tiffany

---

8   The series includes *The Wee Free Men* (2003), *A Hat Full of Sky* (2004), *Wintersmith* (2006), *I Shall Wear Midnight* (2010) and *The Shepherd's Crown* (2015).

and the Chalk, the piece of land she lives on, cares about and protects against danger and neglect.

In the series, the witch's power comes from her connection to the land and its community, which extends from the practical position as a resolver of peoples' disputes to the mystical role of the chosen protector of the land. In return, Tiffany draws from this connection strength beyond her human capacities when facing danger. Nevertheless, her power is not based on commanding nature but on merging with it. As I have argued elsewhere (Dědinová 2022), the witch in Terry Pratchett's fictional world stands as a mediator between culture and nature.

The Chalk is a living whole encompassing grass-covered hills, inhabited by human and nonhuman animals. It can be understood as a fictional representation of the mesh, emphasising not only the interconnectedness and dynamics of the individual components – both material and immaterial – but also a time scale far beyond the span of human life. The very name Tiffany refers in the old language of the story to deep time: it means Land Under Wave and recalls the gradual emergence of the Chalk from the shells of tiny organisms which once inhabited the long-gone sea.

Moreover, the Chalk is endowed with consciousness and agency and chooses the witch as a mediator and protector with whom it sometimes communicates directly, pouring its memories into the human mind. The witch and the land provide each other with meaning and purpose: "She belonged to the Chalk. Every day she'd told the hills what they were. Every day they'd told her who she was" (Pratchett, *A Hat Full of Sky* chap. 5). Tiffany and the Chalk are, to some extent, inseparable, subject and environment co-creating each other in their interdependence:

> The Chalk was her world. She walked on it every day. She could feel its ancient life under her feet. The land was in her bones, just as Granny Aching had said. It was in her name, too; in the old language of the Nac Mac Feegle, her name sounded like "Land Under Wave," and in the eye of her mind she'd walked in those deep prehistoric seas when the Chalk had been formed, in a million-year rain made of the shells of tiny creatures. She trod a land made of life, and breathed it in, and listened to it, and thought its thoughts for it. (*A Hat Full of Sky* chap. 2)

Conceivably, the most expressive metaphor of the link between humans and nature appears in *A Hat Full of Sky* (2004), when a noncorporeal being occupies Tiffany's body, and her soul seeks refuge in the innermost sanctuary in her mind. This final refuge takes the form of the Chalk downlands, manifesting the permeable boundary between Tiffany and her land. As she battles the hiver, the grassy hills take the shape of Tiffany's body and drive the hiver from her. A witness of the struggle asks whether the witch dreams of the hills or the hills dream of being the witch (*A Hat Full of Sky* chap. 8). The answer would be – both. The Chalk and Tiffany are, in effect, one extended entity: a hybrid land-human creature bridging the human–nature divide. Tiffany literally carries the hills in her bones (chap. 3) and the soul of the land in her head (chap. 8). In the fictional world of the series, this is where the deepest source of magic lies, in the mutual dependency and interconnection.

The anthropomorphism of nature and natural entities is common in fantasy literature, often blurring the distinction between anthropomorphic belief and interaction. In fantasy worlds, anthropomorphic belief must manifest differently from the real world and the texts of mimetic literature. While in our world, the idea of a vengeful rain or an obsessed moon pursuing its object of passion is a manifestation of anthropomorphic belief and a considerable anthropo- (or outright ego-) centrism, in fantasy, such things are entirely possible. For the hybrid characters in Pratchett's texts, what is primary in this respect is not the boundary between anthropomorphic interaction and belief but the extent to which anthropomorphism is a platform for the emergence of a relationship, mutual engagement and communication, as theorised by Airenti or Caracciolo.

In the Tiffany Aching series, the Chalk is a living hyperorganism interacting directly with its witch. Humans are part of the land that has its agency, and Tiffany can draw power beyond human limits from it because of her intimate connection with the land. The culture–nature gap is minimised through hybrid characters, yet humans do not lose their autonomy. Tiffany does not physically merge with the land; she remains in a human body, her connection with the world of nature is a matter of choice as well as both rational and emotional acceptance of responsibility for it. Tiffany turns to nature to draw both psychological and magical strength and solace from it, and her connection to the land also gives her a sense of purpose in caring for the Chalk and its human and nonhuman inhabitants. By endowing the Chalk with agency and a mentor-like role, Pratchett uses the consolation of nature as a literary device to bridge the fantasy genre with environmental concerns. In the pairing of Tiffany and the Chalk in the series, the two domains of the metaphor connecting human being and landscape are modified by each other and illuminated from different perspectives, which emphasises their shared origins, connections on many levels, and the permeability of the boundaries between them.

7   Conclusion

"This is not our world with trees in it. It's a world of trees, where humans have just arrived" (Powers), writes Patricia, the fictive representation of forest ecologist Suzanne Simard, in her seminal book. Like the character of Patricia in *The Overstory*, the texts analysed above draw the reader's attention to the interconnectedness of culture and nature, the interdependence and co-creation of humans and their environment, opposing the notion of human exceptionalism and entitlement. While their methods differ, anthropomorphism of nature as a disanthropocentric and mesh-consciousness-building strategy remains essential.

In the works analysed, we can observe a direct relation between the anthropomorphism of nature, or rather the interactional kind of anthropomorphism, and the mesh-consciousness. This implies that characters representing an anthropocentric attitude do not anthropomorphise nature as a whole or its representatives (trees, animals). On the contrary, they adopt a dualistic stance, objectifying nature, perceiving its representatives only as resources to exploit, as inert matter, not as living beings, however different from humans. Examples are the characters in Bacigalupi's "The People of Sand and Slag" and the loggers in Powers's *The Overstory*. While in the latter text, a contrasting approach is represented by characters capable of and willing to engage in anthropomorphic interaction, this direct contrast is absent in Bacigalupi's novella. Bacigalupi's characters fail to move beyond anthropocentrism in their relationship with the world and can be understood as a stark warning against consequences of human exceptionalism.

In the works of Powers and Pratchett, the nonhuman world is primarily portrayed as a living entity or entities that possess their own agency. In *The Overstory* and the Tiffany Aching series, trees and the Chalk are not only cherished by the main characters, but they also actively establish mutually beneficial relationships with humans. The personhood of both human and nonhuman agents emerges from their structural coupling. It is not only humans who change their surroundings – trees and land have similar effects on human characters, shifting, empowering, and influencing them. Human characters are to a large extent defined by their connections to the nonhuman world, which is already reflected at the level of their names (Tiffany in Pratchett or the environmental activists in Powers). Anthropomorphic interactions between human and nonhuman entities allow for a mutual approximation, highlighting bonds and similarities that do not reduce the nonhuman to something more human-like (as would be the case with the isomorphic projection typical of anthropomorphic belief) but enrich our perception of both interacting parties. Humans and nonhumans approach each other but retain their autonomy and uniqueness. Simultaneously, the anthropomorphic interaction effectively highlights the diversity and contingency of interconnections, spanning different spatial and temporal scales, between the human and the nonhuman within the mesh.

The three selected works each, in their own specific way, highlight the dependence of humans on the natural world not only to satisfy physical needs but as a fundamental psychological necessity. Just as the trees in Powers' novel are depicted as entities connected in communities through innumerable bonds, humans are made whole precisely through their relationship with the natural world. While Bacigalupi's novella exemplifies an apocalyptic narrative where losing connection with nature leads to dehumanisation, *The Overstory* and the Tiffany Aching series foreground the restorative and consoling potential of the nonhuman nature. These texts invite readers to perceive nature not solely as a source of grief and loss but also as a wellspring of hope, solace, and meaning, thereby fostering a more positive and constructive environmental ethic. The texts chosen for analysis in this chapter illustrate the problematic nature of the notion of man's separateness from nature and remind the reader of the vitality, fragility as well as the intrinsic value of nonhuman life represented in the chosen texts by animals, plants and natural elements. They also highlight the restorative potential inherent in the nonhuman. Furthermore, as the studies cited above demonstrate, a relationship with the natural leads to acceptance of responsibility for the mesh of which human beings are a part, and a willingness to advocate for the betterment of the planet on which both human and nonhuman life depends.

Bibliography

Airenti, Gabriella. "The Development of Anthropomorphism in Interaction: Intersubjectivity, Imagination, and Theory of Mind." *The Cognitive Underpinnings of Anthropomorphism*, edited by Gabriella Airenti, Marco Cruciani and Alessio Plebe, Frontiers Media, 2019, pp. 7–19.

Alaimo, Stacy. *Bodily Natures: Science, Environment, and the Material Self.* Indiana University Press, 2010.

Arias-Maldonado, Manuel. *Environment and Society: Socionatural Relations in the Anthropocene.* Springer, 2015.

Attenborough, David. "Climate Summit Glasgow Speech Transcript." *Rev.com*, 12 November 2021, https://rev.com/blog/transcripts/david-attenborough-cop26-climate-summit-glasgow-speech-transcript. Accessed 30 Nov. 2021.

Bacigalupi, Paolo. "The People of Sand and Slag." *Pump Six and Other Stories*. E-book, Night Shade Books, 2008.

Bennett, Jane. *Vibrant Matter: A Political Ecology of Things*. Duke University Press, 2010.

Caracciolo, Marco. "Strange Birds and Uncertain Futures in Anthropocene Fiction." *Green Letters*, vol. 24, no. 2, 2020, pp. 125–139. doi: 10.1080/14688417.2020.1771608.

Caracciolo, Marco. *Narrating the Mesh: Form and Story in the Anthropocene*. University of Virginia Press, 2021.

Clark, Timothy. *The Cambridge Introduction to Literature and the Environment*. Cambridge University Press, 2011.

Clark, Timothy. "Nature, Post Nature." *The Cambridge Companion to Literature and the Environment*, edited by Louise Westling, Cambridge University Press, 2014, pp. 75–89.

Clark, Timothy. *Ecocriticism on the Edge: The Anthropocene as a Threshold Concept*. Bloomsbury, 2015.

Clark, Timothy. *The Value of Ecocriticism*. Cambridge University Press, 2019.

Dědinová, Tereza. "Embodying the Permaculture Story: Terry Pratchett's Tiffany Aching Series." *Fantasy and Myth in the Anthropocene*, edited by Marek Oziewicz, Brian Attebery and Tereza Dědinová, Bloomsbury Academic, 2022, pp. 74–87.

DeLanda, Manuel. *Assemblage Theory*. Edinburgh University Press, 2016.

Ganea, Patricia, Caitlin F. Canfield, Kadria Simons-Ghafari, and Tommy Chou. "Do Cavies Talk? The Effect of Anthropomorphic Picture Books on Children's Knowledge about *Animals*." *Frontiers in Psychology*, vol. 5, 2014, pp. 1421–1433. doi: 10.3389/fpsyg.2014.00283.

Garrard, Greg. *Ecocriticism*. Routledge, 2004.

Grasso, Chiara, Christian Lenzi, Siobhan Isa Speiran, and Federica Pirrone. "Anthropomorphized Nonhuman Animals in Mass Media and Their Influence on Human Attitudes Toward Wildlife." *Society & Animals*, 2020, pp. 1–25. doi: 10.1163/15685306-BJA10021.

Haraway, Donna. *When Species Meet*. University of Minnesota Press, 2007.

Heise, Ursula. *Sense of Place and Sense of Planet: The Environmental Imagination of the Global*. Oxford University Press, 2008.

Iovino, Serenella, and Serpil Oppermann. *Material Ecocriticism*. Indiana University Press, 2014.

Kukkonen, Karin, and Marco Caracciolo. "Introduction: What Is the 'Second Generation?'" *Style*, vol. 48, no. 3, 2014, pp. 261–274. doi: 10.5325/style.48.3.0261.

Milton, Kay. *Loving Nature: Towards an Ecology of Emotion*. Routledge, 2002.

Moore, Bryan L. *Ecology and Literature: Ecocentric Personification from Antiquity to the Twenty-first Century*. Palgrave Macmillan, 2008.

Morelle, Rebecca. "Mariana Trench: Deepest-ever sub dive finds plastic bag." *bbc.com*, 13 May 2019, https://bbc.com/news/science-environment-48230157. Accessed 10 June 2022.

Morton, Timothy. *Ecology Without Nature: Rethinking Environmental Aesthetics*. Harvard University Press, 2007.

Morton, Timothy. *The Ecological Thought*. Harvard University Press, 2010.

Powers, Richard. *The Overstory*. E-book, W. W. Norton & Company, 2018.

Pratchett, Terry. *A Hat Full of Sky*. E-book, Doubleday, 2004.

Servais, Véronique. "Anthropomorphism in Human-Animal Interactions: A Pragmatist View." *The Cognitive Underpinnings of Anthropomorphism*, edited by Gabriella Airenti, Marco Cruciani and Alessio Plebe, Frontiers Media, 2019, pp. 36–45.

Tam, Kim-Pong. "Anthropomorphism of Nature, Environmental Guilt, and Pro-Environmental Behavior." *Sustainability*, vol. 11, no. 19, 2019, pp. 1–19. doi: 10.3390/su11195430.

Tam, Kim-Pong, Sau-Lai Lee, and Melody Manchi Chao. "Saving Mr. Nature: Anthropomorphism Enhances Connectedness to and Protectiveness toward Nature." *Journal of Experimental Social Psychology*, vol. 49, no. 3, 2013, pp. 514–521. doi: 10.1016/J.JESP.2013.02.001.

Weik von Mossner, Alexa. "Affect, Emotion, and Ecocriticism." *Ecozon@: European Journal of Literature, Culture, and Environment*, vol. 11, no. 2, 2020, pp. 128–136. doi: 10.37536/ECOZONA.2020.11.2.3510.

PART 2

Elemental Entanglements as a Source of Personal Consolation

∵

Soil Solace in New Nature Writing: Coming to Terms with Transience in Elizabeth-Jane Burnett's *The Grassling*

Bożena Kucała

Abstract

This chapter analyses the consolation of nature in *The Grassling* (2019) by Elizabeth-Jane Burnett, a contemporary English-Kenyan poet, academic and environmental activist. It is argued that whereas Burnett's "creative non-fiction" is representative of the emergent genre of new nature writing as defined, among others, by Robert Macfarlane (2013), and in particular its "nature cure" strand, it also bears marks of the author's deeply subjective approach. *The Grassling* describes the writer's experience of coming to terms with her ailing father's decline and imminent death. Faced with the prospect of an inevitable parting from him, Burnett reinforces the deep bond they share primarily by rebuilding her connection with the place in which his family are rooted. She explores the Devonshire landscape, its wildlife and its history in order to become an integral part of her parent's native environment. A vital part of her quest for a consolatory, transformative experience is her attempt to communicate and integrate with the more-than-human through sensory, embodied interaction, deliberate relinquishment of her anthropocentric perspective and the development of a new language for interspecies communication. The author's topophilia becomes the inspiration for an all-encompassing, time-bending perspective. In accordance with its subtitle "A Geological Memoir," *The Grassling* foregrounds the significance of soil as the foundation for continuity and endurance. Ultimately, the acceptance of individual life as a part of the ongoing vital cycle within a larger-than-human time scale enables the writer to reconcile with the prospect of transience and death.

Keywords

new nature writing – nature in literature – nature cure – autobiography – Elizabeth-Jane Burnett

Elizabeth-Jane Burnett's *The Grassling* (2019) stemmed both from the author's personal experience and her professional and writerly fascinations. The subtitle "A Geological Memoir" reflects the duality of Burnett's motivation. Prompted by her father's imminent death, Burnett composed *The Grassling* as a record of her protracted farewell to him over the months leading to his passing. On the other hand, Burnett's need to express her grief and come to terms with the inevitable converged with her preoccupation with nature as an ecopoet and an activist. An antecedent of *The Grassling* was Burnett's first publication called *Swims* (2017), a book of poetry based on the author's own experience. A creative writer and an enthusiast of wild swimming, in *Swims* she depicts "div[ing] into open water across England and Wales, plunging into rivers, lakes and seas in a watery circuit that takes in the Ouse, the Teign, the Channel, Grasmere and King's Cross Pond in London" (Leah). The writer insists that her academic work, creative writing, environmental activism and private commitments are inextricably intertwined.[1] In an interview with Richard Leah following the publication of *Swims*, she asserted: "I'd been thinking through swimming as helping in activism, but it gets a bit more personal in that sequence. It's more about helping you through other things as well, like mourning" (Leah).

Burnett's multifaceted, partly personal motivation has resulted in works which may appear idiosyncratic or "whimsical," as Clare Saxby says in her review of *The Grassling* (32),[2] but which in fact closely correspond to the emerging convention commonly known as "new nature writing." In the words of Robert Macfarlane, himself its well-known practitioner,[3] at the beginning of the twenty-first century Britain "is going through a golden age of nature writing" ("New Words" 166). If this writing has a hallmark, it is likely to be its generic indeterminacy. Macfarlane describes it as "[r]agtag, wayward and polymorphous." Incorporating aspects of "memoir, travel, ecology, botany, zoology, topography, geology, folklore, literary criticism, psychogeography,

---

1 Elizabeth-Jane Burnett is an English-Kenyan poet and academic. In 2017 she published *A Social Biography of Contemporary Innovative Poetry Communities*. 2021 saw the publication of her second volume of poems, *Of Sea*, described by a reviewer as "closely observed nature poetry about tiny, briny beasts" (Saunders).
2 Apart from *The Grassling*, Burnett's commitment to the fields of her native Devon gave rise to another project, in collaboration with visual artist Rebecca Thomas, in 2017. Their shared interest was in the connections between geology and social history, the millions of organisms below the ground and the human traces that the ground contains. The result of the project was an artists' book in the *leporello* format, with Burnett's text and Thomas's images (Tucker 137–139).
3 Macfarlane is known as the author of, among others, *Mountains of the Mind: A History of Fascination* (2003), *The Wild Places* (2007) and *The Old Ways: A Journey on Foot* (2012).

anthropology, conservation and even fiction," new nature writing typically employs an amalgam of discourses, ranging from the poetic to the scientific ("New Words" 166–167). It has also been somewhat oxymoronically categorised as "creative non-fiction" (Clark 5), in which, typically, factual, objective accounts are intertwined with poetically evocative, first-person narratives. However, the number of works that exemplify this new tendency is perhaps sufficient to regard it as a new genre in its own right that eschews the existing categories. In 2008 the *Granta* magazine devoted to it an issue titled "The New Nature Writing," which, according to Alexander J. B. Hampton, may be viewed as an attempt to mark the beginning of this movement (455).

Deploring the relative lack of recognition of Burnett's book, which she regards as "seminal" to contemporary nature writing, Xiaoxiao Ma argues that its value "lies in its brave, innovative experiment in the entanglements of sentiment and knowledge, language and landscape, family history and natural history, poetry and prose" (431). This chapter argues that, while relying strongly on the parameters of new nature writing, *The Grassling* is also marked by the author's deeply felt individual experience of loss and transience. Specifically, the memoir is rooted in Burnett's bond with her father. The father's native Devonshire village of Ide and its countryside appear to be extensions of himself, therefore exploring the area is an effective way for the protagonist to maintain and strengthen her connection with him. Hence, besides universalising reflections, *The Grassling* is concerned with a particular place and a particular individual, just as the search for a meaningful bond with nature, typical for new nature writers, is suffused with the author's unique approach – another amalgam of the conventional and the specific is Burnett's celebration of soil,[4] both as the substance that constitutes the earth, and the constitutive

---

4 As a natural resource that sustains life on earth, soil has been crucial to the history of mankind because it shapes the material environment of human habitation (cf. M. R. Balks and D. Zabowski, *Celebrating Soil: Discovering Soils and Landscapes* [2016]; Richard D. Bardgett, *Earth Matters: How Soil Underlies Civilization* [2016]). However, it also has inspired a variety of cultural practices, beliefs and rituals. For example, it is often seen as a component of individual and collective identity. As Edmund Chapman notes, "Along with blood, soil is one of the most common tropes in conceptualizations of national belonging" (51). This approach is well-rooted in the German context. Martin Heidegger's concept of *Bodenständigkeit* (groundedness) illustrates "a wider understanding of the (national) soil as guaranteeing the self and/ as part of the *Volk*" (Chapman 51). The focus on "blood and soil" was fundamental to the Nazi definition of national identity. In *Soil and Culture* (2010) Edward L. Landa and Christian Feller argue that "soils touch people's lives on a variety of levels – from the intellectual, to the pragmatic, to the spiritual" (xix). Their book presents diverse perspectives on soil, including its representations in literature and the visual arts. Another wide-ranging discussion of the cultural significance of soil may be found in Nikola Patzel et al.'s *Cultural Understandings*

stuff of her and her ancestors' native Devon. It is by turning to the source and sustenance of all life that she ultimately finds comfort and consolation.[5]

1   New Nature Writing as a New Convention

It is a paradox in a book which describes the author's attempt to erase the boundary between her self and nature that it should be mediated and shaped by texts. Burnett is well aware of the literary provenance of her work. In one of her conversations with her father she identifies her project as "nature writing" and reassures him that "this has become a popular genre" (*The Grassling* 137).[6]

Don Scheese traces the beginnings of modern nature writing back to the 1800s, "when romantic writers confronted the forces of modern industrializa-tion" (4). He regards it as a modern manifestation of the antiquated pastoral dream of "escape to" or "quest for" rural harmony and simplicity, sustained by the conception of nature as a retreat or sanctuary (4, 6). Typically, early nature writing took the form of a first-person account of an exploration of a mostly non-human environment, customarily carried out on two levels: out-ward, physical as well as inward and mental or spiritual (Scheese 6). Argua-bly, Henry David Thoreau's writings such as *Walden* (1854) or *Walking* (1862) epitomise the most influential nineteenth-century type of such narrative. However, it would be a gross oversimplification to trace a single line of con-tinuity between the nineteenth-century models of nature writing and con-temporary works. The history of nature writing has been shaped by different literary and cultural traditions, in addition to different natural environments. And so whereas American nature writing abounds in images of the wilderness, its British counterpart, rooted in the legacy of Romanticism (cf. Rigby), tends to focus on domesticated landscapes (Westling 2–3). British nature writing of the first half of the twentieth century was largely dominated by an apprecia-tion of the rural landscape as a site of continuity, tranquillity, and an anchor

---

of Soils (2023), which explores "societal and individual, philosophical, religious, spiritual, psychological, political and educational aspects" of "humanity's relationship with soil" (12).

5   Another project Burnett was recently involved in was designed to celebrate the uplifting power of nature. Members of the public were asked to contribute to a poem commending the coming of spring in 2021. Burnett gave the final shape to the crowdsourced poem, which she called "Spring, An Inventory." In the words of Stephen Morris, she wanted it to be "an optimistic riposte to the grim statistics – deaths, Covid-19 cases, hospital admissions – that have been such a feature of the past 12 months" (Morris).

6   In *The Grassling*, she alludes to two classics of nature writing: Rachel Carson's *Silent Spring* (1962) and W. G. Hoskins's *The Making of the English Landscape* (1955).

for national identity (J. Smith 38). For example, in his very popular book *In Search of England* (1927) H. V. Morton cherishes the English countryside as a comforting alternative to the daunting, alien modern world (J. Smith 38). Henry Williamson's *An Anthology of Modern Nature Writing* (1936) contains comparable instances of "mythification and insular retreat" (J. Smith 39). Yet the idealisation of nature and the idea of rural idyll were even then criticised and dismissed as outdated or hackneyed (Gifford 53; Hampton 454–455) and, especially in the post-war years, as irrelevant to and oblivious of the contested political and social problems of modernity (Gifford 52–53). By contrast, contemporary writing has become conspicuously responsive to environmental issues – accounts of authors' uplifting communion with nature are frequently interspersed with their concern with pollution, extinction, climate change, the human encroachment on the natural. Nevertheless, Terry Gifford argues that the pastoral ideal of retreat into nature is not dead, but, in a modified version (as "post-pastoral") continues to inspire contemporary works as well. Drawing on the examples of several British and Irish poets, Gifford demonstrates that they innovatively explore the human–nature connection while engaging with environmental ethics and cultural geography (63). Deborah Lilley comes to similar conclusions concerning today's British nature writing: in its twenty-first-century incarnations, it continues to rely on "familiar features of the British tradition" while being sensitive to diverse and changing meanings of nature, especially with regard to the current ecological crisis (Lilley [4]).

It appears that the traditional opposition between the city and the country has not been eroded either. In his 1973 study *The Country and the City* Raymond Williams questioned both of the stereotyped images of rural and urban life in English literature and writers' tendency to juxtapose them. Yet in numerous narratives representative of twenty-first-century nature writing, it is still the world of nature where individuals turn in order to find solace, consolation and inner harmony, leaving behind the turmoil and pressures of the city. For example, in *The Outrun* (2016) Amy Liptrot tells the story of her mental recovery in the wilderness of the Orkney Islands, after a decade of living in London, struggling with addiction and a series of personal crises.

As Samantha Walton contends in *Everybody Needs Beauty: In Search of the Nature Cure* (2021), the ubiquity of the current search for "the nature cure" may be treated as, on the one hand, the resurgence of a deep-rooted primal human need, and, on the other hand, as a fashion liable to trivial institutionalisation and commercial exploitation (2). Testing the restorative potential of nature in her double role, as an individual in need of healing and as an academic who scrutinises this widespread tendency from a detached, informed perspective, Walton takes issue with superficial, naïve gestures of reconnection with the natural world. She is also sceptical about the dichotomy between the city and the pristine natural world out there: "The old definition of 'nature' as a world separate from humanity may not mean much in a world of microplastics, urban sprawl and climate change" (4). Nonetheless, both her private escape into nature in search of health and her scholarly inquiry into the viability of "the nature cure" originate at the same moment as she escapes from the city, which she describes as "the first step in a ritual that's at once staggeringly ancient, and absolutely modern" (2).

Elizabeth-Jane Burnett prefers to uphold the traditional division; indeed, it would appear that complete, instinctive immersion in nature is what she regards as necessary in the process of overcoming her crisis. Her "geological memoir" is a personal testimony to the effectiveness of "the nature cure." *The Grassling* emphasises the separation between the country and the city, although the city is conspicuous by its absence. The Devon countryside is hardly pristine wilderness,[7] and yet it offers the narrator the chance of an encounter with the natural in contrast to the urbanity of her daily life. Without overtly vilifying the city, Burnett opposes the two spheres and presents the city as a place of transitory, superficial living, where one is unable to put down roots:

> Here, in the field, we've used our own mulch, grown our own vegetables, fruit, flowers, trees. But I can't remember doing much in the years I lived in gardenless flats in London. Now, in Birmingham, although there's a garden, as a rented property I don't have free rein over it. I wonder what the figures are on this; people disconnected from the soil because of lives too transient to grow attached to it. (Burnett 58–59)

In the editorial introduction to the above mentioned *Granta* issue, Jason Cowley observes that the authors writing in the new mode "don't simply want to walk into the wild, to rhapsodize and commune: they aspire to see with a scientific eye and write with literary effect" (9). Burnett certainly does it all, yet the wish for restorative engagement with the other-than-human gains prominence. In her view, it is in communion with nature that one can experience different time scales and different modes of being, that the self can undergo expansion and transformation, whereas the city appears to thwart such processes: "It is only when I am in a distant city that I lose my elasticity," claims the author of *The Grassling* (172). Yet, apart from occasional remarks such as this, her professional, urban life is not represented at all, which shifts the focus of

---

7   The complexity of the notion of "wilderness" in the British context is discussed, among others, by Olga Roebuck in *The Place It Was Done* (147–149).

her book wholly onto an account of her escapist forays into the rural countryside, from which she emerges restored and revived. When challenged by her
father about her environmental activism, Burnett prefers to shun the subject
(*"Please don't ask me anything about climate change"* [57]), instead upholding
idealised visions of rural life as an instance of harmonious human–nature
co-existence and collaboration (e.g. "Principled and independent, these old
farming families have given, as they have taken, from the soil" [155], which she
describes as "a helpful human stewardship of the soil" [114]).

If new nature writing sometimes looks like a sentimental return to a naïve
celebration of the wild or the rural, it also draws on the discourse of science –
"the disembodied, objective and impassive voice of the expert," represented by
biologists and conservationists (Hampton 455). Such a jarring combination of
the factual and detached on the one hand and, on the other hand, celebratory
and reverential approaches to nature can be observed in *The Grassling* as well.
For instance, while the author explains that birds navigate thanks to a protein
in their eyes which enables them to detect magnetic fields (71), she also ponders the mystery of their migration, supposedly analogous to the obscure and
irresistible "pull" that takes her away from her desk, her room, "in a building, in
a city, miles away," towards her father's land:

All the skin of me [...] is called to the soil. All the water of me, the churning
motion, is called to the fish [...]. All the heart of me, the pulsing inside, is
called to the pulsing outside: the grass, the birds, the insects, the worms;
and further, to the beating that continues beyond the earth, beyond anything tangible: to my father's fathers. (73)

The manifestly metaphysical echoes in Burnett's reflections recur throughout her book, and again resonate with much of the mainstream nature writing today. In "Post-secular Nature and the New Nature Writing" Alexander J. B.
Hampton argues that the new writing has inherited the Romantic "unease with
the construction of the rationalized relationship of the individual and nature"
(457), legitimising a subjective, intuitive response. While it avoids an open
engagement with a religious framework, this writing still tentatively enters the
realm of the sacred by destabilising the "sacred–secular dichotomies" (460).
Thus, it navigates a way between an accurate, knowledge-based description
and an appreciation of the enchanted, mysterious and awe-inspiring dimension of the natural world that escapes human control and understanding.

While actively seeking to break down the barrier between the human and
the more-than-human, the narrator of *The Grassling* is self-conscious about
her transitions between the two realms. During her solitary walks in Devon,

she experiences privileged moments of oneness with the soil and the wildlife surrounding her. And yet, such deeply sensuous, instinctive experiences are occasionally impeded by her human consciousness reasserting its autonomy and separateness from the natural world. For instance, a prolonged episode of complete immersion and dissolution of the self, described in the chapter "Yslende," comes to an end when reality begins to reassume its normal shape for the narrator and she tries to rationalise what has just happened: "I blink. The field moves back to wherever it has been, no longer bound by gold but green. I start to question what I've seen, where the map has taken me" (171). There is an element of deliberate re-enchantment and re-sacralisation of nature in the author's quest for communion with the natural world. Yet, unlike Wordsworth, who, on his solitary walk, was involuntarily astounded by the sight of the daffodils and the vision of cosmic harmony that they revealed, Burnett actively strives to find the concord and unity that she believes are out there. There may be an allusion to Wordsworth in the chapter "Daffodil," in which Burnett describes taking a bunch of flowers to her father. The daffodils complete the "list of everything yellow" that she has compiled on her journey (20, 23). The colour appears to impose unity on all things, pervading them with light and joy. In a characteristic mix of registers, she associates yellow with "the new meanings" that arise from combinations of blocks of chemicals, with colour therapy, and with "a magic dust" and "a gust of calm" that can be released in her father's room (21).[8]

Indeed, the assumption that appears to underpin *The Grassling* is that whereas modern civilisation may have estranged us from nature, the rift may be healed because a primal bond with our more-than-human ancestors can be rediscovered due to the shared genesis in the eternal soil. Burnett's own evolutionary history supposedly derives from "the original bedrock of Devon" (124). Contemplating approvingly the rhythms of rural life and her family's rootedness in the human- and natural history-laden landscape, the narrator speculates that "[t]he Burnetts, perhaps, 300 million years ago, were reptiles, adapting to a new life solely on land" (124). This, in turn, enables her communication with other beings: "And as the wood pricks into the clay, sometimes smoothly, at other times having to pick its way among the grains, it says: look. This is the soil that you come from. This is the soil of your fathers" (125).

---

8   Kate Rigby detects a convergence between Goethe's meditation on the sensual-ethical impact of colours and Wordsworth's depiction of his response to the daffodils. In his *Theory of Colour* (1810) Goethe describes the effect of bright yellow as "pleasurable and cheering, with an element of vivacity and nobility in the force with which it works" (qtd. in Rigby 56).

Burnett's depiction of human entanglements in the natural world may be related to the concept of *sympoiesis*, widely deployed in contemporary biological sciences. Derived from Greek, the term literally means "making-with" (Haraway 58). Originally coined by Friedrich Schlegel (Rigby 18), the concept was taken up by Donna Haraway to refer to dynamic systems in which "[c]ritters [...] make each other through semiotic material involution, out of the beings of previous such entanglements" (Haraway 60). As she explains, "It is a word for worlding-with, in company" (58). Implicitly echoing the notion of *sympoiesis*, Burnett's descriptions of nature invariably emphasise the organic unity and interconnections between all animate and inanimate natural entities, which she tries to convey in a style typical of new nature writing: "descriptive proliferation, a kind of lush prose or thick description" (cf. Hampton 465). In the words of Timothy C. Baker, the world that *The Grassling* projects is "one of complete entanglement" (120). Burnett's accounts try to dissolve the divide between humans, plants and animals, stressing their shared vitality and interdependence in the eternal cycle. In her interactions with the wild, she clearly tries to achieve a relational sense of selfhood (cf. Crowther 51–52). While encountering other species, the author reminds herself to discard her human perspective and appreciate nature on its own terms. Thus, a comment about an owl's "unnatural" craning of the head is immediately self-corrected, because the movement supposedly appears unnatural only to humans, while Burnett recognises that the otherness of non-human species "shakes us out of our usual scripts" (47).

The author's effort to experience the world of nature causes her to focus on her bodily experiences, but, further, also to transcend the limits of her own corporeality and imagine what it is like to have a different material form and thus to experience the world differently. Her experiments in overcoming the mind–body dualism chime with the philosophical stance of embodied cognition. Its basic tenet is that

[t]he properties of an organism's body limit or constrain the concepts an organism can acquire. That is, the concepts by which an organism understands its environment depend on the nature of its body in such a way that differently embodied organisms would understand their environments differently. (Shapiro and Spaulding)

In *The Mind Incarnate* Laurence A. Shapiro argues that perceptual processes, such as vision or audition, are "tailored to bodily structures." For instance, human vision depends on having two eyes and being able to move the body in a certain way; organisms with non-human bodies "will have nonhuman visual and auditory psychologies" (Shapiro 190). A more radical claim within embodied mind theories is the notion of "embodied thought," which states that, besides the generally recognised social factors, the body has a prominent role also in the content of our thoughts and our conception of the world (Shapiro 191).

In *The Grassling*, embodiment plays a crucial role in the narrator's interaction with her environment, not as "a reification of selfhood, but a form of connection with the land," as Timothy C. Baker notes (120). The narrator's efforts to adopt a non-human, or more-than-human perspective frequently entail a physical change of posture and a more sensuous, palpable contact with the wild. And so she touches the bark of trees, bends low to make her way through the entangled vegetation, immerses herself in the local rivers, lies on the forest floor, soaks in the sun, smells the scent of wet earth, lets herself be scrutinised by deer and birds, and conjures up an image of herself as the titular grassling growing in the soil. In such "elemental return to earthy origins" the protagonist "imagines living as a true inhabitant of the Earth" (P. D. Smith). Her encounters with the more-than-human can certainly be described as "transformative" (T. Baker 120). Burnett's purpose is to share her material existence with other creatures as an embodied being, therefore she cherishes moments when a sense of connectedness prevails, when she feels a part of a living system larger than individuals and larger than the human world: "A moth flutters by my fingers; we all match perfectly and seem synecdoches of each other: the dusky drab looper moth, my tawny hand, the rose-brown pine needle floor" (63).

In "Tintern Abbey," Wordsworth speaks of feeling "a presence," "a motion and a spirit" that "rolls through all things" (207). In her discussion of the legacy of the Romantic contemplative ecopoetics, Kate Rigby suggests that from a secular perspective the religious experience of oneness with the universe may be reinterpreted as a sense of "intimate enmeshment," or "enfleshment," connecting the animate and the inanimate, the living and the dead, "human and otherwise" (35). Without attempting to name what she exalts, the narrator of *The Grassling* has a sense of a living force that pervades and unites all things: "The blood in the antlers, the sap in the tree, the moonlight in me – running. The trill of the stream, the turn of the roots, the slap of the hooves – running. The blood in me starting to slow, soil settling below, shoulder blades finally letting go" (42). Yet, while Wordsworth is led to "elevated thoughts," Burnett's numinous experiences remain distinctly corporeal and sensory – as she tries to absorb natural things with her senses and her whole body, she strives after reciprocity. The unique, nearly mystical moments of merging with the world of nature invariably involve an experience of virtually Ovidian metamorphosis. Lying in a field of wheat, the narrator feels "sap" circulating in her body, her "petals" pulling apart, her skin "ripening" as she seems to grow into the soil: "I am a long way into turning into something else: from animal to grass, from grass to hay [...]." It is in moments such as this that "[e]verything responds, everything is on: listening and being heard are simultaneous. The sounds of becoming and belonging are the same as the spaces opened up for them" (170). Like other contemporary ecologists, Burnett appears to believe that "'*out there*' is a different world, older and greater and deeper by far than ours" (cf. Campbell 130).

Discussing paradigmatic narratives of retreat from social life to nature, such as those by Henry David Thoreau or his twentieth-century follower Annie Dillard,[9] Randall Roorda claims that they are tales "not of adventure or diversion but rather of absorption." Intrinsic to them is a desire for a kind of "conversion" – "an element of action, quest, pursuit of the extraordinary and transformative" (10).[10] Burnett follows – literally and mentally – in the footsteps of these and other walkers, seeking experiences that will help her find lasting peace and achieve a perspective from which loss, mortality and transience become acceptable. The process involves a surrender of the rational human grasp of reality, and the embodied experience of opening herself to elemental immersion. As a result, the individual, rational, time-bound self is transcended. In keeping with much new nature writing, Burnett's narrative is punctuated by passages that describe "re-wilding the self and re-integrating it into nature" (Hampton 461). As Margaret Kerr and David Key argue in "The Ecology of the Unconscious," contact with wild places "changes our sense of self. 'I' as a 'part' becomes different because of the whole" (65).

However, the pursuit of immersion in natural processes and recounting the experience seem to be incompatible goals because the narrator inevitably confronts the problem of finding an adequate language in which to express the apparently inexpressible. In *Landmarks* (2015), Robert Macfarlane comments on nature's resistance to articulation: "Nature does not name itself. Granite does not self-identify as igneous. Light has no grammar. Language is always late for its subject" (10). Burnett's awareness of the difficulty inherent in her project is signalled by two of the epigraphs to *The Grassling*. One, taken from Monica Gagliano's "Plant Communication" (2018), refers to the scientific study

---

9   Dillard is best known as the author of *Pilgrim at Tinker Creek* (1974), a work often compared with *Walden*.

10   However, although generally appreciated by ecocritics, the tradition of nature writing which prioritises the author's psychological or spiritual experiences has also been criticised for supposedly evading "ecological and environmental issues," which, according to Dana Phillips, "are biologically, socially, and politically as well as, if not rather than, psychologically and spiritually determined" (203).

of the sound waves emitted by plants. In the other excerpt, from Enid Blyton's
The Magic Faraway Tree (1943), the possibility of talking to plants is legitimised
by the fairy-tale convention. In the vein of new nature writing, Burnett seeks
both a means of communicating with the more-than-human and a way to
transcribe this communication through her repeated efforts to transcend the
boundaries between species and between elements. With her perception of
the natural world clearly derived from the idea of an intertwined ecosystem
rather than a hierarchy of species, the author assumes that language is not
an exclusively human property. In her view, other beings are endowed with
a form of speech too, which, however, remains inaccessible to her: "As I run,
I am conscious of not speaking the language, of the weight I give the grass and
the ground below, the pressure on the soil. [...] I don't even know who I'm not
speaking to. What is grass? What is in and under it?" (61)[11]

Nevertheless, the narrator resumes her efforts to break down the barrier.
Like the character in Blyton's story she references, Burnett tries talking to trees
and imagines their response (27–28). A more effective strategy is trying to
imaginatively shed her human form and become a member of another species:
"As I stand in the trees, I am one" (45). Moments of imaginary metamorphosis
correspond to the author's attempts to devise a new language to breach the gap
between herself and other beings. Chapter 7, "Grass Diaries," is written in lyrical, visually-charged disjointed sentences which record a dialogue between the
protagonist and a blade of grass. The grass watches, listens, responds. However,
to reverse this (clearly unwanted) effect of personification, in several other
chapters the narrator portrays herself as a grassling, growing out of the soil and
experiencing the natural environment through its non-human body. The short,
elliptical, impressionistic sentences in which those chapters are narrated illustrate Burnett's endeavour to convey a different perception of the world through
a different kind of language, to infuse "the notion of oneness with the soil"
into the narrative itself (Saxby 32). The writer deliberately blurs the identity of
the narrator – the "Grassling" chapters are predominantly written in an impersonal style, which impedes the identification of the point of view. In contrast
to the rest of the book, those chapters avoid the first-person pronoun, referring to the Grassling as "it." The Grassling's language is rooted in the soil and
organically related to it: "Living in the soil is thick and raw, and the Grassling
claws in the hair of words. The grass's speech, the birds' high beep, rabbits'
low feet, churning" (96); "If grass has a voice, it is fast and scorns punctuation"

---

11   Burnett's meditations bring to mind Walt Whitman's celebration of the mystery of
     the grass in *Leaves of Grass*: "A child said *What is the grass?* fetching it to me with full
     hands" (53).

(95). The fragments of poetry inserted in the narrative illustrate the quality of this voice by blurring divisions between words. In a quasi-imitation of the evolution of speech, the Grassling painstakingly learns to produce sounds and subsequently seeks its own sound by listening to and trying out the sounds of other creatures. Yet this is a twofold process – the search for individuality is underscored by the acknowledgement of one's place in a plurality of voices: "that is exactly what a field is: millions of throats like this, millions of passages; channelling air and water and movement and things constantly touching and coming apart; it is all this, every blade of it" (142).

The idea that human language is mimetically related to the world of nature is also advanced in the more realistic, or less idiosyncratic, parts of *The Grassling*. Guided by her father's book of local history,[12] Burnett traces the origins of contemporary place names to their Anglo-Saxon or Celtic meanings, anxious to prove that there is an iconic match between the features of the landscape and their names in human speech. *The Grassling*'s working title was "A Dictionary of the Soil" (143)[13] – this intention is still reflected in the structure of the book, which consists of chapters arranged alphabetically, according to their one-word titles. But the entire book may be seen as the fulfilment of the author's ambition to be a mouthpiece for the soil: "as I look across the wide expanse of field and hill, out to the shine of the Exe and the distant sea, I think about the language of this place, of myself, and how I must keep it open. How I must say out, out in grass voice, in field voice, in open-throated hill voice" (159). After her father's death, the symbolic act of writing the names of her family on the grass as she is about to return to the city is yet another attempt to connect human language and human existence with the primal elements.

2   Soil Solace – an Authorial Mark on Nature Writing

It is clear that Burnett continues the ancient and modern tradition of eulogising nature as a site of retreat, consolation and spiritual comfort. A prominent strand in contemporary new nature writing are hybrid texts in which, in the words of Harriet Baker, "nature writing meets recovery memoir" (33) – the

---

12   Donald Burnett, *A History of the People and Parish of Ide* (1992).

13   Burnett's interest in words denoting the human relations with nature resembles Robert Macfarlane's project which resulted in the publication of *Landmarks*. Macfarlane scoured the British Isles in order to compile a glossary of local, often forgotten words describing natural phenomena, or what he calls "land language" – "mutual relations of place, word and spirit" ("The Word-hoard").

writer's engagement with the natural world takes place in the context of his or her strife to overcome a personal crisis. As Richard Mabey notes, "The idea of 'nature cure' goes back as far as written history. If you expose yourself to the healing currents of the outdoors, the theory goes, your ill-health will be rinsed away" (223). His own book, *Nature Cure* (2005), recounts the author's explorations of the flatlands in East Anglia as part of his ultimately successful attempt to overcome depression. His process of recovery is permeated with a desire to reconnect with fellow humans and with the natural world. The writer stresses the need to find a balance "between nature and culture" in his life (226). As Mabey reflects in hindsight, whereas his new romantic relationship, devotion to work and appreciation of elements of the world of nature were steps towards healing, ultimately it was his sense of unity with the other than self that enabled him to overcome his crisis: "it was regaining that imaginative relationship with the world beyond that was my 'nature cure'" (64).

Personal, healing prose is represented by a host of other contemporary narratives, such as Olivia Laing's account in *To the River* (2011) of her walks along the River Ouse in Sussex after the break-up of a relationship, Katharine Norbury's *The Fish Ladder* (2015) – a narrative of the author's consoling walks along various rivers in the wake of a miscarriage and the loss of family members, or Amy Liptrot's memoir of curing herself from an addiction in the wilderness of the Orkneys (*The Outrun* [2016]). In *Moth Snowstorm: Nature and Joy* (2015) Michael McCarthy draws on his deep connection with nature to deal with his childhood trauma of losing his family home. Walking in her late mother's footsteps in the mountains of her native Scotland was Sarah Jane Douglas's way of coping with her recent loss, as described in *Just Another Mountain* (2019). A well-known example of "nature cure" writing is Helen Macdonald's *H is for Hawk* (2014) – a memoir of the year she spent training a hawk, which helped her come to terms with her father's death. All these are works in which nature provides the space where a different perspective is gained and new meanings are discovered (cf. Hampton 456).

Preparing for her father's passing – which indeed occurred during the process of writing – was the main motivation behind Burnett's *The Grassling*. As can be inferred from the above examples, "nature cure" narratives customarily focus on a chosen element of the natural world. In *The Grassling*, Burnett accords special importance to soil as the foundation of all life. By exploring the particular place connected with her father's family, she both learns to reconnect with him, and also, conversely, to let go of her beloved parent. Accordingly, her musings on soil are derived both from her affection for the local land and her appreciation of the universality of the earth. During one of her first exploratory walks in the countryside near Ide, she reflects: "The magnetism of the land, not just where I was birthed, but where my father was; his father and his; pulls me to it, as if by knowing it, I should know them" (15). Her father's book of local history as well as conversations with him enable her to perceive the land as a palimpsest preserving traces of numerous generations of the same family. In parallel to her visceral response to the soil, in the course of her research she goes deep into the layers of the human history in her native Devonshire village, as far back as the medieval Anglo-Saxons and Normans. The meaning of a family tree (Chapter 6, "Family Tree") is both literal and metaphorical – the author is curious about the memory of trees and about her family's rootedness in their native place. Her tendency to erase the boundaries between human and more-than-human beings, between human and natural history is manifest here as well. She is aware of treading the paths that her ancestors have walked, and of walking along the roots of ancient trees (26).

As an instance of what Hampton terms "topophilic" literature[14] within the emergent genre of new nature writing, *The Grassling* emphasises the "complexity and depth of a single place," interweaving nature and culture, natural history with human history, folklore and etymology (Hampton 456). There is undoubtedly a nostalgic strain in Burnett's appreciation of the disappearing farming traditions and local crafts in Devon that she comes across during her walks. Her reveries intimate that whereas she sees urban life as disconnected from the nourishing influence of nature, in her view rural life has always been smoothly aligned with the landscape, the wildlife and the cycle of the seasons. Even the old railway bridge that she comes upon during her perambulations, rather than standing out as a human artefact, appears to blend with the environment: "Here is something that men built. That living bone and sinew hauled and handled. A disused space carries its past with it, does not entirely give up its use value after it's discarded. I see the men that built its bridge and railway, as much as I see the stone before it" (107–108). In contrast to the city, an epitome of rootless existence, living in close proximity to the natural environment entails a continuity with the generations of one's family, who, like the Burnetts, have inhabited the same place for hundreds of years. The author's exploration of the village of Ide and its vicinity includes searching the local archives for traces of her ancestors. The archives, however, cover only recent history. Nevertheless, she is convinced that the soil compensates for deficiencies in human memory and historical records as it continues to connect the living and the nameless dead: "When the archives run out of documents and

---

14   In his seminal book *Topophilia: A Study of Environmental Perception, Attitudes, and Values* Yi-Fu Tuan defines the term as "the affective bond between people and place or setting" (4).

people with local knowledge pass away, only the land remains. And as even that recedes, will it seem fanciful that there were ever these interlocked lives working and growing in the soil?" (81)

In a strand of new nature writing such as Burnett's, which focuses intensely on a chosen locality, the preoccupation with the particular, as Neal Alexander put it, "establishes the basis for much more expansive metaphysical speculations in which the local and the cosmological are imaginatively conjoined" (4). The centrifugal orientation observable in much topophilic writing is enabled by the recognition of "humanity's embeddedness in nature" (Hampton 456). The "expansive" quality of *The Grassling* is a vital contributor to the consolatory effect. For the writer, acquiring a different, more distanced perspective is an essential component in the process of reconciling herself to individual death. If viewed within a broader framework, transience inevitably appears to be a universal course that all living beings undergo. Accordingly, Burnett's approach consists in the expansion of the temporal context of her observations so that her father's life and death are subsumed within the succession of generations, and beyond that, within the daunting eons of geological history. From a sufficiently detached point of view, her father's eighty years are only a small contribution to the story of the human settlement in this land, and merely an infinitesimal fraction in the history of the soil. She asks rhetorically: "What is two thousand years in the span of the soil?" (15). The author places human history in the context of what John McPhee called "deep time" in his 1981 book *Basin and Range*, with the implication that the geological perspective makes human affairs seem irrelevant (cf. Macfarlane, Introduction xɪ). Yet what Burnett seeks by acknowledging the co-existence of diverse time scales is not so much quiet resignation and acceptance of individual demise, but the possibility of transformation and a kind of natural afterlife. The author's belief that the primal unity of the Pangaea will turn out to be a cyclical event in the history of the earth nurtures her hope that, analogously, her father may arise again, even if in some unimaginably remote future and in an altered shape: "that he'll come back together, re-form, in a deep culm of earth and time. With no end" (19).

Therefore, the comfort that may be derived from adopting the scale of geological time also is contingent upon the expansion of the concept of what constitutes a living being. Much as she stresses the affinities between humans, plants and animals, Burnett sees even more basic connections – all organisms are made of the same particles circulating in nature. That is why the author of *The Grassling* prioritises soil as the foundation which preserves, breaks down, transforms and nourishes various forms of life: "While there is still so much to learn about the soil, we know that the interaction of organisms in the soil contributes to the cycles that make all life possible" (69). Viewed in the context of these never-ending cycles, simultaneously creative and destructive, the demise of an individual may be reconceptualised as a part of the process of enduring transformations. A body buried in the soil will decompose into nutrients which will subsequently sustain other beings. In her periphrastic, poignant description of her father's burial, Burnett writes: "But this place knows him, and the grass takes him to it like a part of itself returning" (182).

One of the best-known literary expressions of such sentiments may be found in the poetry of Dylan Thomas. In "A Refusal to Mourn the Death, by Fire, of a Child in London" (1945) the poet insists that death, understood as being united with "the first dead" and "the grains beyond age," must not be mourned. "After the first death," he solemnly declares, "there is no other" (101). Yet an opposing attitude may be found in his poem "Do Not Go Gentle into That Good Night" (1951), in which the poet urges his father to resist death: "And you, my father, there on the sad height, [...] Rage, rage against the dying of the light" (116). Neither is Elizabeth-Jane Burnett completely reconciled to the prospect of losing her father. Travelling to Devon and hoping that she will find him still alive, she reflects: "Transformation is a dominant process in the lower horizons, as clay may undergo chemical changes to its structure; or, lower down, soil may be turning from fibric to sapric as organic matter decomposes. But I do not want his transformation. I want a return" (50).

However, such wishes are, of course, unrealistic; dominant in Burnett's narrative is a celebration of the soil as the (literally) tangible realm of continuity and endurance. Her soil-worship takes the form of ritualistic, sensuous gestures: feeling clumps of soil with her hands, tasting it, leaving her footprints on its surface. The book she writes and the rituals she performs in relation to the world of nature are preparations for the final parting from her father. In the last chapter, after he has been committed to the soil, his daughter is ready to let go. The act of writing her father's name and the names of her ancestors on the grass, despite its apparent futility, should not be regarded as an empty gesture, but as an expression of Burnett's recognition that human life is like the grass that grows out of the soil – evanescent and fragile, yet perennial and renewable. Walt Whitman once arrived at a similar recognition in *Leaves of Grass*: "The smallest sprout shows there is really no death" (54).[15] Burnett's final action is a kind of prayer addressed to the soil as the sole site of consolation: "I rest on my knees and call out to my father. To my family. To the ground that gives and takes them" (187). Timothy C. Baker is right to observe that "it is

---

15   In the final section of *Song of Myself* Whitman writes: "I bequeath myself to the dirt to grow from the grass I love" (96).

not a voice of desperation, or nihilism" (125). The project of coming to terms
with her loss by resorting to the primal element of earth has resulted in the
paradox of simultaneous closeness to her father and a universal, distancing
perspective. It is the latter stance that allows the author to finish on a note of
reconciliation: "That great continuing. Glow" (187).

Eventually, in Burnett's "Geological Memoir" writing and communing with
nature converge, and both are revealed to have been ways of forging a bond
with her father: "Through a shared space and shared narrative, I write myself
into him" (149). Even though, as has been argued above, in many respects her
book emulates the emerging convention of new nature writing, the author
has also found her own fulfilment of the potential for consolation inherent
in the world of nature. Essential to the success of her project is her "uniquely
experimental, sensory style" (Saxby 32), which matches her sense of oneness
with the soil. However, the main factor that lends *The Grassling* its distinct-
ness is the noticeable authenticity of the writer's feelings for her father, whose
presence, as Clare Saxby notes, "is palpable on every page" (32). It may be said
that there appears to be a tension in Burnett's narrative between individuation
and a dissolution of selfhood, between, on the one hand, her commitment to
her father and a particular patch of soil, and, on the other hand, her quest for
more-than-human integration with the natural world. These two are, however,
interrelated; in the words of Timothy C. Baker, her care for her father and her
care for the soil are "not simply anthropocentric, but entangled" (122). As the
narrator confesses, the intimate experience of nature opens up "new kinds of
living" to her, enabling her to "live a wider life" (62). Ultimately, the contemplative, immersive ecopoetics proposed by Burnett, grounded in the concept of
creaturely *sympoiesis* and timeless continuity of the cycle of life, provides a
viable framework for dealing with the singularity of personal grief and helps
to outbalance the confrontation with death with a sense of participation and
renewal.

Bibliography

Alexander, Neal. "Theologies of the Wild: Contemporary Landscape Writing." *Journal
    of Modern Literature*, vol. 38, no. 4, 2015, pp. 1–19.
Baker, Harriet. "Orcadian Ego: How a Drinker Was Saved by the Scottish Landscape."
    Times Literary Supplement, June 10, 2016, p. 33.
Baker, Timothy C. *New Forms of Environmental Writing: Gleaning and Fragmentation.*
    Bloomsbury Academic, 2022.

Bubíková, Šarka, and Olga Roebuck. *The Place It Was Done: Location and Community in Contemporary American and British Crime Fiction.* McFarland & Company, 2023.

Burnett, Elizabeth-Jane. *The Grassling.* Allen Lane, 2019.

Campbell, Sue Ellen. "The Land and Language of Desire: Where Deep Ecology and Post-Structuralism Meet." *The Ecocriticism Reader: Landmarks in Literary Ecology,* edited by Cheryll Glotfelty and Harold Fromm, University of Georgia Press, 1996, pp. 124–136.

Chapman, Edmund. "Language, Soil, and 'Jewish' Alienation in Levinas and Adorno." *Diacritics,* vol. 51, no. 1, 2023, pp. 50–73.

Clark, Timothy. *The Cambridge Introduction to Literature and the Environment.* Cambridge University Press, 2011.

Cowley, Jason. "Editor's Letter: The New Nature Writing." *Granta,* vol. 102, 2008, pp. 7–12.

Crowther, Rebecca. *Wellbeing and Self-Transformation in Natural Landscapes.* Palgrave Macmillan, 2019.

Gifford, Terry. "Towards a Post-pastoral View of British Poetry." *The Environmental Tradition in English Literature,* edited by John Parham, Ashgate, 2002, pp. 51–63.

Hampton, Alexander J. B. "Post-secular Nature and the New Nature Writing," *Christianity & Literature,* vol. 67, no. 3, 2018, pp. 454–471.

Haraway, Donna J. *Staying with the Trouble: Making Kin in the Chthulucene.* Duke University Press, 2016.

Kerr, Margaret, and David Key. "The Ecology of the Unconscious." *Vital Signs: Psychological Responses to Ecological Crisis,* edited by Mary-Jayne Rust and Nick Totton, Routledge, Taylor & Francis Group, 2012, pp. 63–77.

Landa, Edward R., and Edward Feller, editors. *Soil and Culture.* Springer, 2010.

Leah, Richard. "Elizabeth-Jane Burnett: 'Swimming Can Give You the Optimism to Keep Going.' Interview." *The Guardian,* 16 November 2017. https://www.theguardian .com/books/2017/nov/16/elizabeth-jane-burnett-swims-interview. Accessed 13 Feb. 2023.

Lilley, Deborah. "New British Nature Writing." *Oxford Handbook Topics in Literature,* online ed., edited by Oxford Handbooks Editorial Board, Oxford Academic, 2013 [pdf pp. 1–18]. doi: 10.1093/oxfordhb/9780199935338.013.155. Accessed 2 Jan. 2024.

Ma, Xiaoxiao. "*The Grassling: A Geological Memoir.*" *Green Letters: Studies in Ecocriticism,* vol. 26, no. 4, 2022, pp. 430–432. doi: 10.1080/14688417.2022.2106684.

Mabey, Richard. *Nature Cure.* 2005. Pimlico, 2006.

Macfarlane, Robert. Introduction. Tim Robinson, *Stones of Aran: Pilgrimage.* Faber and Faber, 2008, pp. IX–XIV.

Macfarlane, Robert. "New Words on the Wild." *Nature,* vol. 498, 2013, pp. 166–167.

Macfarlane, Robert. *Landmarks.* Hamish Hamilton, 2015.

Macfarlane, Robert. "The Word-hoard: Robert Macfarlane on Rewilding Our Language of Landscape." *The Guardian*, 27 February 2015. https://www.theguardian.com /books/2015/feb/27/robert-macfarlane-word-hoard-rewilding-landscape. Accessed 13 Feb. 2023.

Morris, Steven. "'A Surge of Hope': Public Helps Create Poem Celebrating Coming of Spring; Writer Elizabeth-Jane Burnett Brings Together 400 Voices for Optimistic Riposte to Events of Past Year." *The Guardian*, 21 April 2021. https://www.theguardian .com/books/2021/apr/21/public-helps-create-poem-celebrating-coming-of-spring -elizabeth-jane-burnett. Accessed 28 Dec. 2022.

Patzel, Nikola, et al. *Cultural Understandings of Soils: The Importance of Cultural Diversity and of the Inner World*. Springer Nature, 2023.

Phillips, Dana. *The Truth of Ecology: Nature, Culture, and Literature in America*. Oxford University Press, 2003.

Rigby, Kate. *Reclaiming Romanticism: Towards an Ecopoetics of Decolonization*. Bloomsbury Academic, 2021.

Roorda, Randall. *Dramas of Solitude: Narratives of Retreat in American Nature Writing*. State University of New York, 1998.

Saunders, Tristram Fane. "Poem of The Week: Elizabeth-Jane Burnett." *The Daily Telegraph*, 1 May 2021, p. 5. *ProQuest*. Accessed 3 Nov. 2023.

Saxby, Clare. "Dictionary of the Soil. Elizabeth-Jane Burnett, *The Grassling*." *Times Literary Supplement*, 7 June 2019, p. 32.

Scheese, Don. *Nature Writing: The Pastoral Impulse in America*. Routledge, 2002.

Shapiro, Lawrence A. *The Mind Incarnate*. A Bradford Book, 2004.

Shapiro, Lawrence, and Shannon Spaulding. "Embodied Cognition." *Stanford Encyclopedia of Philosophy* (online). 25 June 2021. https://plato.stanford.edu/entries /embodied-cognition/#EcolPsyc. Accessed 13 Feb. 2023.

Smith, Jos. *The New Nature Writing: Rethinking the Literature of Place*. Bloomsbury Academic, 2017.

Smith, P. D. "*The Grassling* by Elizabeth-Jane Burnett Review – A Geological Memoir." *The Guardian*, 21 March 2019, p. 16. *ProQuest*. Accessed 3 Nov. 2023.

Thomas, Dylan. *Collected Poems 1934–1952*. J. M. Dent & Sons, 1957.

Tuan, Yi-Fu. *Topophilia: A Study of Environmental Perception, Attitudes, and Values*. With a New Preface by the Author. Columbia University Press, 1990.

Tucker, Judith. "Walking Backwards. Art Between Places in Twenty-First-Century Britain." *Walking, Landscape and Environment*, edited by David Borthwick, Pippa Marland and Anna Stenning, Routledge, Taylor & Francis Group, 2022, pp. 127–146.

Walton, Samantha. *Everybody Needs Beauty: In Search of the Nature Cure*. Bloomsbury Circus, 2021.

Westling, Louise. Introduction. *The Environmental Tradition in English Literature*, edited by John Parham, Ashgate, 2002, pp. 1–8.

Whitman, Walt. "Song of Myself." *Leaves of Grass*. With an Introduction by Gay Wilson Allen. The New American Library, 1958, pp. 49–96.

Wordsworth, [William] and [Samuel Taylor] Coleridge. *Lyrical Ballads 1798*, edited by H. Littledale, Oxford University Press, 1959.

CHAPTER 7

More Than Backpacks and Swims: British Nature Writing and the Consolatory Role of Nature

Olga Roebuck

Abstract

The chapter focuses on the genre of nature writing and its response to the changing notion of the environment. The two chosen texts, Roger Deakin's *Waterlog* (1999) and Hugh Thomson's *The Green Road into the Trees* (2012), show multiple ways in which the environment plays a consolatory role, predominantly by allowing a more accurate and up-to-date identity definition. Rather than focusing on a subject-centred and uninvolved scientific depiction of the environment, these works show personal engagement and suggest possible strategies of reconnecting and enjoying the transcendental dimension of the interface between the human and the natural. In its initial part, the article argues that the contemporary approach to the environment must change to facilitate such a reconciliation, and new nature writing, as a genre devoted to the interface between the human and the natural, thus changes accordingly. Without shedding its Romantic heritage completely, the analyzed works deepen the genre's insight into, for example, the political dimension of human relationship to the environment. Deakin and Thomson discuss the contrast between the ownership and stewardship of nature, the issue of conservation and its meaningfulness. In the section dealing with immersion and embeddedness, the analysis stresses the importance of experiential knowledge – the experience of physical immersion opens new cultural layers of the environment for both writers, activating their awareness of several collective and individual identities. The final part of the discussion highlights how the writers go beyond the purely conscious awareness of the self and explore the possibilities of transcendental reconnections. The chapter thus combines an analysis of the genre's new scope with an application of critical insights to a classic text of new nature writing (Deakin) and a text that has not yet received substantial critical attention (Thomson).

Keywords

environment – new nature writing – Roger Deakin – Hugh Thomson

Much has been written on both the perception of nature and the development of the genre of British nature writing. Yet, the changing conditions of the Anthropocene require the constant re-assessment of the human relationship to the natural environment, including its literary depiction. This chapter looks at the genre of British nature writing, its development and its future with a particular emphasis on the genre's way of rendering the consolatory role of nature. There are multiple ways in which new nature writing envisions the redefinition of human identity as part of the environment, i.e., the reconciliation of the human and the natural, thus counteracting the current sense of estrangement or alienation from nature. One of the many questions arising at the beginning of the intended analysis is how the contemporary approach to the natural environment must change in order to facilitate such reconciliation. The two chosen texts that provide at least partial answers to the posed questions include Roger Deakin's *Waterlog* and Hugh Thomson's *The Green Road into the Trees*. Deakin's 1999 account of his, to some perhaps eccentric, swimming journey through Britain, has become a must-read among those who love roaming, wandering, and pondering. Thomson's "Walk through England" (citing the book's subtitle), published in 2012, has attracted academic attention but so far only moderate general readership despite it being awarded the Wainwright Prize for Nature and Travel Writing in 2014. Both books help to illustrate that the search for consolation in nature centres around the new possibilities for human self-definition and, perhaps, even signposts a way towards re-defining human–nature relationships. Finding one's new identity as part of the environment defined by the human and the natural interface is a reservoir of solace.

The traditional perception of nature as a resource for humans has changed only very little. The prevailing tone in discussing environmental protection is that of reformed environmentalism. As Timothy Clark summarizes, the mainstream approach still sees the natural world as an economic as well as cultural resource for human beings but stresses the need for sustainable development and protection of aesthetically attractive landscapes (Clark 1). The environmentalist imperative is thereby weakened by the economic strife of the capitalist industrial society, simply requesting moderation rather than full abstinence from environmentally damaging practices. Voices calling for an overall change in the human relationship to the physical environment are labelled as radical because they demand a reassessment of the anthropocentric view of the world. David Kidner stresses the difficulty of changing these traditional approaches because "environmental problems, more than most, have been understood as arising from styles of thought and action that are historically established" (Kidner 61). Even the genre of nature writing, which first yielded ecocritical readings, employs the anthropocentric perspective of imposing meaning on nature and consuming it. But changing the "almost all-pervading assumption that it is only in relation to human beings that anything else has value" (Clark 2) will necessarily have a significant cultural impact. The essentially anthropocentric character of nature writing has been recognized e.g. by Abberley et al. as something almost impossible to escape, as the genre does not locate value in the environment per se, but in the humans' experience of it. Yet, the outcome is clearly in line with ecological politics even if the appreciation of nature includes the worshipping of ourselves: it collapses the human/nature binary (Abberley et al. 23). After all, the central theme of this publication: escape into nature and seeking consolation in nature, are essentially anthropocentric themes, yet with an important environmentalist tone. Refraining from unrestrained consumption and, on the contrary, reconstructing human identity as part of the natural environment may, apart from the consolatory effects on individuals, also benefit the rather troubled genre of nature writing.

There are many problems inherent in the genre of nature writing, the first and foremost being its main theme, nature itself. Once famously called "the most complex word in the language" by Raymond Williams (176), nature has become a contested term. Should Williams be deemed too dated, the same concern over the complexity of nature or the wild can be found in the much more contemporary writer Richard Mabey. In his *Nature Cure* (2005), he characterizes the wild as a "distinctive aesthetic, ecological and moral category, anything that is resistant to human control, prediction or understanding and represents the unmanaged energy of nature" (219). All in all, both the terms "nature" and "the wild" have generally been replaced by a more inclusive concept of the "environment."

The environment is not limited by the quality of being pristine and allows for the inclusion of the so-called edgelands, or landscapes bearing traces of past human activity. Such suggestions take varied forms and result in varied terminology. According to Neal Alexander, some distinguish between the outdated and criticized concept of *wilderness*, expressing scepticism over its almost "sacramental value," and *wildness* which they seek in "the most ordinary overlooked landscapes" (2). These edgelands may offer more varied wildlife than some of the carefully managed natural reserves. Including the edgelands and writing humans into the environment, thereby attempting "a true social history, an up-to-date ecology" (Smith 2), is a path taken by nature writing at the turn of the millennium and followed since. Refraining from the separation of nature and culture also corresponds with Jamie Lorimer's characteristic of the Anthropocene as being "multinatural" (7) and inclusive of human interaction with the many recurring processes resulting in constant flux. Lorimer offers the perception of the environment as "the commons, the everyday affective site of interconnection between the human/non-human entanglement in which the human relinquishes some of his or her detached superiority" (11). Thus, the environment is approached as a living thing which cannot be depicted as a static object of admiration or protection.

Another problem of nature writing lies in the general presumption that it has the tendency to view the environment romantically as a retreat, refuge, or place of redemption, or that it resorts to the scientific description with a conservationist tone. It is thus often dismissed as escapist and as lacking a political commitment. Although the genre's firm roots in popular culture may give grounds for such a judgement, many examples prove that nature writing reaches towards new and experimental forms and reflects contemporary challenges. The dilemma of whether its rendering of nature is inclined towards the rhapsodic or whether the genre is to remain in the scientific non-fictional domain has recently been resolved as a combination of both.

Although some authors dispute the label of *new nature writing*, it has significant traces of the original Romantic tradition as well as placing an emphasis on accurate factual descriptions inherited from natural history. Unlike the original forms of nature writing, however, its novelty stems from the central position of the author and his/her unique rendering of the environment in this complex genre. Although critical points have been raised against the centrality of this authorial voice, emphasizing mainly the anthropocentric implications it may have for the genre (Alexander 3), I see this subjectivity as a sign of a personalized dialogue between the human and the non-human. New nature writing thus, in the words of Alexander Hampton, does not strive to be "a monological imposition that pretends that the eye of the beholder offers an objective rendering of the observed natural world" (458). It is a subjective, individual, and dynamic rendering because it occurs at the interface between nature and culture.

There are nature writers whose works follow the path suggested by Richard Mabey in his book *Flora Britannica*, which Smith characterizes as a "true social history, an up-to-date ecology of plants and human beings" (Smith 2). One of them is Roger Deakin and his *Waterlog* (1999), which Clark sees as having a partly non-human focus and suggests that even his "metaphorical statements have a cognitive value of some ecologically valid kind" (8). His swim through the British Isles is not only a true immersion in various local landscapes but also a deep analysis of local natural and social conditions – created, sustained, or preserved by mutual interaction between nature and humanity. Another example can be found in Hugh Thomson's *The Green Road into the Trees* (2013).

Thomson's engaging walk through Britain follows various roads leading far into
the distant past without overlooking the natural and cultural variety offered
by the present. Like Deakin's, Thomson's book balances the artifice and the
scientific depiction to make an ecological statement in a way relatable to the
reader. Their writing shows the environmentalist concern by recognizing, as
Mabey puts it, "both the kindredness and the otherness of the natural world"
(*Oxford Book of Nature Writing* VII). Both works, among other matters, remove
the nationalist landscape aesthetic of the modernist sense from the agenda of
new nature writing and replace it with a more accessible and relevant localism
by recognizing the multi-faceted identity of the local landscape in the complexity of its local histories, art, customs, and natural conditions.

1   Escape without Escapism

Both works take the form of a journey undertaken as if to reconnect with the
environment, searching for one's place within it. Deakin is somewhat straightforward: "To swim is to experience what it was like before you were born" (3).
Where Deakin is literally immersing or plunging in water, Thomson's immersion is more metaphorical: "So I feel like plunging in – and to do so by the
darker, underground ways, again like a mole, tracking the older paths into the
country" (2). Neither of them actually distinguishes between the country and
the environment – confronted by what they see, they try to make sense of and
reconnect with the individual locations, which surprise them with their complexity and variety.

    Both authors employ an openly subjective point of view, which allows them
to reflect upon issues that make their work politically engaged. The minor
episodes, in which they often humorously depict their encounters with local
landscapes, characters and histories, often lead towards much deeper reflections of political and social issues. Deakin's dive into water often collides with
British land ownership and the much-disputed right to unlimited access to
open country. In the chapter entitled "Lords of the Fly," he skilfully juxtaposes
two contrasting approaches to the environment. On the one hand, he presents the Houghton Fishing Club as a remarkable example of environmental
stewardship – a duty to care for the river which co-creates the human community around it. On the other hand, there is the encounter with a guard preventing Deakin from trespassing on the river in Winchester, which is believed to be
private property.

    The opposing approach to the river, as part of open country, is foreshad-
owed already in the description of both locations. Stockbridge, the home of

the Houghton Fishing Club, is characterized as "a rural Venice," apparently trying to make the most of its water and sharing it with the casual passers-by: "How marvellous to find a place that values, uses and enjoys its river like this, instead of tucking it away out of sight" (Deakin 19). Similar local pride in the river is not paralleled in Winchester, where Deakin had hoped to reconnect with a long tradition of swimming in the Itchen, only to be met with *Private – No Access* notices from the very beginning.

In Stockbridge, Deakin is not only plunging into the clear waters of the River Test but also into the meticulously recorded history of the Houghton Club. Its chronicle, enriched by Deakin's commentary, reveals a complex local identity moulded around the natural course of the river and the variety of natural phenomena affecting not only fly-fishing but also local lives around the Test. While Deakin enjoys swimming in the club waters, yet worries about trespassing, his encounter with the locals surprises him by its apparent casualness: "A romantic-looking couple in their sixties passed by through the meadow and we exchanged a polite 'good afternoon'. They did their best to look unsurprised" (20). Deakin portrays this location and its approach to the environment as a true interface between nature and culture. Common history, local language, and literature reflecting upon the plethora of social relations all refer to the proximity of the river.

The author's lush prose and enchantment with the environmental stewardship of Stockbridge portrays the acknowledged "sensitivity and awareness of place, describes a familiar landscape rendered afresh with the unfamiliar, including human geography of history, folklore and etymology, interwoven with natural history" (Hampton 456). However, Deakin employs a rather different note in his rendering of Winchester. Referring to Winchester's rich history of public swimming in the river and the abundant literary mentions of local beauty, he seeks to immerse himself in this tradition as well as in the waters of the Itchen. Yet the whole place emanates an air of possession and ownership, and passers-by are met with requests to be quiet, as university exams are in progress, or not to trespass. Deakin employs a gradually stronger critical voice with ironic undertones. These culminate upon his encounter with a porter, the College River Keeper, and an Alsatian, all rather nastily demanding him to leave, for the river is private property. For Deakin, this episode represents a springboard for disputing the tradition of ownership in Britain, allowing the privileged few to limit other people's access to open country and thus, to significantly restrain their freedom to roam. As Marianna Dudley suggests, "formulation of place may dominate the ways in which we are able to engage with it" (49). Limiting access to any part of the open country (including rivers) thus denies environmental knowledge of the place to various groups of people, which has significant ecological consequences. After all, Dudley suggests that recreational engagement with a place creates highly nuanced environmental knowledge (49) and as such plays a crucial role in environmental protection.

Deakin's *Waterlog* proves the above: the author's political statements are strengthened by frequent mentions of ecological issues which he combines with the criticism of land ownership and social privilege. He comments on invasive species threatening local ones, fish farming and its negative impact on local habitats, the unhealthy corruption of many conservationist societies or authorities, etc. Far from the disengaged, purely conservationist style, Deakin's social criticism shows how new nature writing has become political. He stresses the importance of experiential knowledge being available to everybody in order to make ecological concerns public. It is precisely because the environment is understood as the interface between the human and non-human, between nature and culture, that allows for the beneficial effects of reconnecting with nature.

Hugh Thomson's *The Green Road into the Trees* was published not long after Robert Macfarlane's *The Old Ways* (2012) and, for that reason, may invite inevitable comparisons. The most obvious is that, unlike Macfarlane, Thomson denies that the journey is an escape. James Attlee's review published in *The Independent* makes that difference clear from the very subtitle of the article: "Cross-country adventures of the wanderer and the escapee" (Attlee). When accounting for his journey as an attempt to understand his home country, he also presents this endeavour as an addiction and himself as an incurable wanderer. Likening himself to an alcoholic who cures his hangover with another drink, Thomson does not even unpack his belongings after a previous journey before setting off on his search for rural England, the exploration of which should allow him to come to understand the changes that he notices everywhere and to reconnect with his home. After all, a clear sense of one's own separate identity is a prerequisite for a mature relationship, one that recognizes the difference between oneself and the other while respecting the other (Kidner 66). Therefore, Thomson's initial impulse for the journey is the complete opposite of an escape: he desires to belong to his local community again, to redefine and update his local identity. His starting point is in the estrangement from his environment, whereas the end of his journey brings him the consolation of reconnection.

Thomson's travel is much less solitary than Deakin's. While Deakin focuses a lot on his physical experience which leads him to contemplate more general ideas, Thomson deliberately populates his account with very specific characters. He always presents their names and life stories, thereby creating a set of case studies representing a wide variety of local identities. Ranging from poachers, local conservationists, travellers, farmers, old friends, and many more, the characters lay out their own understanding of rural lives, which Thomson ponders and comments on. Such an approach gives the whole book the quality of a report and makes it truly authentic. Yet, Thomson does not stay above the text as an objective presenter: he gets involved and contrasts the individual stories with his own experience and worldview. The reader can thus actually witness the gradual process of how the author's local identity develops and adjusts according to his experience, constituting the true authenticity of the work.

A journey peopled with such a variety of characters is bound to steer Thomson's attention towards many current issues. Because he attaches these issues to the individual characters and makes them part of their story, they become personal and thus more urgent. Like Deakin, Thomson, for example, addresses the issue of access to the open country, i.e., the right to roam. He illustrates it with the story of Danny, who travels the Icknield Way in his handmade caravan in the opposite direction to the one described in the book.

On one hand, Thomson makes the issue his own (as he argues for the right to access the environment freely), on the other hand, the locals often connect Danny with a controversial community of New Age travellers. The beauty of the traditional caravan is sometimes attractive to the locals as a reminder of the past, and Danny admits that he is invited to camp on village greens, where he clearly serves as a village feature. But when he stays longer than a night, his presence becomes a cause for concern. Other times, Danny's identification with the Traveller community provokes open hostility, which Thomson disapproves of. "Like the good burghers of Suffolk, we like the idea that there are still some romantic souls travelling the highways and byways of old England in the traditional manner. We just don't want them to come our way" (Thomson 180). The aggressive reaction of the local farmer recounted in Thomson's book encapsulates the general fear of otherness combined with the traditional approach to land ownership and the exclusiveness of accessing the open country, which is a sad reality of contemporary rural Britain.

Thomson also frequently addresses the issue of conservation. He contrasts the experience of people who have lived in rural areas all their lives and worked in close connection to the natural environment with clearly defined conservation schemes. His approach mirrors the dichotomy mentioned in the introduction to this text: the pragmatic and Anthropocentric approach to the conservation of places or species that are attractive and *deserve* to be protected, and the contrasting approach characterized as a sustainable cohabitation of the human and the natural. When travelling across Salisbury Plain, Thomson again laments the limited access, because "the army have carved a great deal out of it for firing ranges, and if the 'red flags' are up, you can't walk across" (73). This time, however, the main point is elsewhere. The firing ranges overlap with areas where the bustard bird protection scheme runs. The clearly well-supported and well-financed scheme focuses on reintroducing the bustard into the British countryside at all costs. The futility of the scheme is enhanced by the author's suggestion of the true reason for the demise of the bird: its habit of not nesting but leaving eggs freely on the ground. The subsidy distributed to the farmers in the area for not ploughing their land is juxtaposed with the fires caused by the ordnance from the army firing grounds, which the author watches accompanied by one of the conservationists. Thomson further contrasts the Anthropocentric attempt to manage the environment with a gentle reminder voiced by a local retired gamekeeper, who represents the viable human/natural approach, and whose main concern is the decreasing number of songbirds caused by the protection of predators and raptors. The resulting message resonates with the above-mentioned importance of experiential knowledge of the environment to make meaningful and sustainable decisions about it.

Thus, both writers show no intention of seeking out romantic solitude and escaping from human society. On the contrary, they explore the history of human interaction with the environment and carefully handle its traces, comparing them to today's situation and seeking new possibilities for such future interactions. Through their uplifting interactions with nature, many of the characters that the authors depict help to identify possibilities of a well-balanced co-existence.

2   Immersion and Embeddedness

The importance of the experiential knowledge of a place has already been mentioned in connection with the notion of environmental stewardship. Yet, there is another level on which the physical experience of the environment of a particular place is crucial. Hampton sees the physical immersion in nature as a form of "rewilding the self and reintegrating into nature" (461). Deakin declares such an immersion into the environment from the very beginning as a completely natural desire, as if plunging in water satisfied his "visceral need to return to an elemental part of himself" (Hampton 461–462). Such a declaration not only emphasizes the subjective point of view, i.e., the author's personal need to satisfy his yearning, but it generalizes it as a form of the universal need of the whole of the human species to reintegrate with the environment from which they have become alienated. While Deakin acknowledges

John Cheever's short story "The Swimmer" as one of the main impulses for his swimming endeavour, where being in the water is understood as the return to a natural condition of the protagonist, he only takes that as a starting point. In his own words, he sets out to test whether "for the best part of a year, the water would become my natural habitat" (5) and whether such experience could be extended into a new type of knowledge learned from the non-human world.

Deakin stresses the universality of his environmental reintegration. His ability to share the delights of clear water with various species and articulate that delight in literature can be understood as an example of *symbolic embeddedness*. It is a term that David Kidner presents as a key element of environmental ethics and a necessary component of environmentalism. He characterizes the human re-immersion in the environment as the possibility to regain the lost articulacy of the forgotten "other layers of selfhood" and to voice the non-rational modes of relation between the human and the non-human, undermining the paradigm of the person as a detached, rational subject facing an object world (Kidner 79). It is a task that Deakin undertakes and, ultimately, admits its burdensome accomplishment. His ultimate physical tiredness and the shedding of his belongings on the beach add to the author's apparent relief: "I had left my rucksack and clothes beside a beautiful pebble starfish on the beach, another echo of the Scilly Maze. Perhaps I had at last swum my way through it" (Deakin 331). Deakin's connoisseurship for wild swimming should thus not be dismissed as a subject-centred description of an experience that sets the self above the environment, but quite the contrary: as an articulation of the need to belong and a journey towards reintegration, i.e., re-immersion. It fully complies with Kidner's concept of differentiated and autonomous personhood that is fully diffused in cultural and natural contexts. As such, it may lead to the development of human potentialities and yet be fully consistent with the natural order (Kidner 61). Deakin's publication thus signposts a new way of relating the self to the natural environment and environmental ethics.

For Deakin, swimming represents an important element of his identity. It proves to be a collective one – as he shares a kind of "tribal" experience with a family of swimmers as depicted in the chapter entitled "Tribal Swimming." The story of Judith and her family inhabiting an old mill and enjoying a truly amphibian life shows the power of one's environment. Wild swimming, against which the local Environment Agency warns Judith in an official letter, is a defining feature of the family's local identity. And because "the joys of swimming are sometimes those of silence and solitude, sometimes of communion with nature, and sometimes the more friends, who join you, the merrier" (115), they are willing to share it with Deakin. By joining the tribe of swimmers, he reconnects with all who share the joy of this unique experiential knowledge of their environment.

Thomson is as compulsive in his immersion as Deakin, only he plunges into the intricacies of the British rural environment. His journey represents an attempt to come to understand his own country again (after his extensive travels abroad). In his opening chapter, Thomson, jet-lagged from his last trip, describes a celebration in his local market town that leaves him totally puzzled about his own local identity. His immersion is motivated by the need for reconciliation with his environment. As Kidner asserts, "only in the industrialized world is individuality viewed in terms of nonconformity to cultural mythologies and traditions – an assumption that is central in weakening the self, making it vulnerable to commercial pressures and narcissistic fashions, and assuring that our relation to the rest of the natural world, by default, will be one of uncomprehending exploitation" (72). In many ways, Thomson's journey validates this claim: he seeks to reconnect with cultural traditions and reformulate his relationship with the natural world.

The path that Thomson chooses for the reconnection with his local identity is a historic one. The author believes that he can only come to understand the contemporary changes if he decodes the past of his environment: "If I made a journey, as well as being an investigation of the deepest past, I wanted to explore what was happening now" (Thomson 6). Through this decision, Thomson unwittingly justified Kidner's understanding of environmental theory as an "immersion within a symbolic realm that precedes consciousness, therefore – environmental theory embodies, articulates, and legitimates cultural forms" (Kidner 61). The symbolism of local landscapes speaks clearly to Thomson. He often invokes the notion of rediscovering the meaning of natural/cultural symbols that he had forgotten. When walking along Grim's Dyke in the Chilterns, the bluebells prevent him from pondering the Saxon resistance to the Vikings. This natural symbol of southern rural England which the author had admittedly forgotten about suddenly starts making a deeper sense: "The carpet of blue flowers managed to be a celebration both to the transience of spring and of the permanence of the English landscape" (Thomson 10). The historical continuity of the dyke, now a permanent feature of the local landscape, connected with the fleeting beauty of the bluebells that, however, also "had taken centuries to establish" (Thomson 11), begin to speak to the author as legitimate cultural forms. As the episode comes at the very beginning of his journey, it is possible to state that his immersion into the rural landscape is already starting to yield results.

There is a point in Thomson's account where he shares the same amphibious experience as Deakin. In the middle of his journey, he happens to pass through his hometown and enjoys a stopover. To reflect upon his old habits, he decides to swim back home. He explains the decision as practical as well as emotional, but it again corresponds with his initial attempted reconnection. Like Deakin, Thomson not only immerses himself in water but also in the environment and begins to articulate its selfhood, reflecting upon its forgotten layers. In one paragraph, his style even begins to resemble Deakin's, especially when he indulges in almost natural lyricism full of surprised cows, overhanging chestnut tree branches, or startled kingfishers. On the other hand, he goes beyond a simple description. By swimming home rather than using the usual means of transport to visit his hometown, Thomson realizes the centrality of the place's Anglo-Saxon heritage. Applying his historical approach to analysing the nature of today, he begins to look at the local landscape through a new lens and he can begin to appreciate the Anglo-Saxon culture "not of the courts but of the fields: the farms, homesteads and hunting lodges that had established a pattern of village life which had endured for over a millennium" (Thomson 130). Kidner asserts that "our experience of the world can be understood to have arisen conjointly with our self-experience [...] and most environmental writing has focused on the former while neglecting the potentially exploitative implications of modern subjectivity" (61). Thomson (and Deakin likewise) prove him wrong because the new nature writing makes much more of these exploitative implications. Thomson can use the experience of the world as a springboard for his self-experience to yield a new sense of belonging. In the above-cited case, for example, he realizes that he can re-examine the English village life with a "shocking familiarity" (131). Thomson is able to combine his conscious intelligence with symbolic awareness, which, as Kinder states, "exists within an integrated cosmos in which the particular constantly reminds us of the whole, and the whole finds expression through the particular" (70). Thomson's immersion re-connects him to the environment through the rediscovered historical layers, the cultural symbolism of which begins to speak to him in familiar words.

Both authors set off on an exploratory journey: Deakin aims to examine the effect of physical immersion in water on his own self, and Thomson hopes to find a new understanding of rural England. For both authors, their immersion enriches their selfhood with the consolation of being reintegrated into the environment.

3   Transcendental Reconnection

New nature writing's potential to reach beyond or above the scientific or material description of the environment has already been recognized. Although the genre was at first burdened by its ambition to evade nostalgic depiction and, as such, nature writing approached the sterility of scientific non-fiction, the new nature writers are highlighting the deep structures and networks that address the transcendental. As Hampton suggests, new nature writing works with an enchanted environment which to some extent resists exact description and articulation, and thus provokes "descriptive proliferation, a kind of thick description, oftentimes evoked by these moments of resistance" (465). After all, the purely instrumental and rational relation to the natural world which does not allow for the inclusion of the unconscious or the transcendental, leads to a distanced, destructive, and non-dynamic frame of reference. Hampton goes on to characterize the new nature writing's relationship to the environment as *post-secular* because it challenges "a subject-centred, immanence-bound, disenchanted representation of nature that sets the self over and above nature" (455). Kidner adds, in a similar manner, that "arational awareness can be expressed only when selfhood is recognized to be simultaneously conscious and unconscious when conscious autonomy is balanced by rootedness in an unconscious context in which individuality becomes secondary to immersion in the whole" (69). The apparent need for rediscovered transcendentality of the environment resonates through the work of both Deakin and Thomson.

For Deakin, it is the act of swimming itself which has a transcendental quality. He confesses that flowing with water opens new possibilities as he is convinced that "following water, flowing with it, would be a way of getting under the skin of things, of learning something new" (3). By likening swimming to the act of being born, Deakin presents it as an important initiation ritual which leads to a new type of knowledge. Contrasted with the well-informed reality of our everyday lives, Deakin longs for mysterious, unknown places, the meaning of which he can make himself, through his immersion.

Thomson travels through several sacred spaces or larger sacred landscapes. He admires them as manifestations of the human ability to co-create the environment. The prehistoric sites of worship have become a permanent human imprint on the natural landscape, often indistinguishable from it. His interpretation sees the sacred landscapes partly as an answer to the primal need to stamp, permanently mark, or own the land, and partly as a symbol of a transcendental connection to the environment. The numerous links drawn between archaeological excavations, literary depictions, public beliefs, and Thomson's experiential knowledge of sacred sites, create a mysterious and complex transcendental experience, often drawing a large question mark at the contemporary treatment of these places.

Throughout his journey, the author keeps reminding us of our lack of understanding of sacred landscapes, including the best-known ones such as Stonehenge. This inability to understand, i.e., our disconnection from prehistoric sacred landscapes has a historical background, often believed to start with the Roman colonization of Britain and only deepened later by the Christian fear of everything pagan. However, Thomson sees the problem as a disconnection not only from one's history but also from the landscape and the transcendental perception of the environment, typical for prehistoric civilizations in different parts of the world. He compares Salisbury Plain to the sacred landscapes of the Andes, where meaning is given "to the harsh and difficult environment to create a complex sacred landscape" (69). Our inability to read the meaning of these ritual sites today is thus directly caused by our inability to understand them in the context of the surrounding landscape.

Ritual or sacred landscapes also do not speak to most contemporary visitors due to our predominantly pragmatic and material approach to the environment, as discussed above. Alexander characterizes our preoccupation with a landscape in a dual sense: "as the material environments inhabited by human and non-human beings and as a way of seeing spaces and places that is closely linked to a pictorial or textual representation" (2). This material preoccupation is clearly reflected in the museum-like treatment of sacred landscapes. Thomson reflects upon several obvious misconceptions, Stonehenge again being one of them. On one hand, there are clear signs of neglect and lack of funding; on the other, the stones were fenced off in the late 1970s and have remained inaccessible since. "This," according to Thomson, "provides fine distant views of the stones but is hardly an immersive experience" (63). The author's deliberate violation, as he enters the site, proves the transcendental experience that the other visitors are denied.

A similar set of misconceptions surrounds a similar ritual landscape at the end of Thomson's journey, Seahenge. The dispute over the removal and museum display of its parts attracted significant attention. Even though Thomson admits that the central inverted oak removed from the circle of Seahenge "retains all its power, even behind glass" (280), it is a significant hindrance to the experiential knowledge needed to appreciate this sacred landscape. More generally, treating sacred landscapes as a fixed entity or a museum exhibit goes against the need to acknowledge and incorporate the "symbolic and metaphorical dimensions of experience that inhabit a more meditative, trancelike state of mind allowing us to move toward a more accurate empathy with the natural world" (Kidner 73). Thomson's account of his own meditative experiences in several sacred landscapes of rural England serves as a good example.

It is a meditative experience that leads Thomson at the end of the journey to the final formulation of his sought identity. Although the book represents a thorough examination of rural England's environment through many cultural phenomena, because culture is, after all, understood as "an essential medium that roots us into the world" (Kidner 63), the final revelation of the core of his identity comes to him at Seahenge. This monument to temporality marks the volatile borderline between the sea and land, and Thomson extends this volatility to the whole notion of English identity. "Rather than hold on to some outmoded notion of national identity, like a piece of the driftwood out in the ocean, we should just let go and have the waves take us where they will" (Thomson 284). This beautiful simile, which includes the environment in our identity formation, acknowledges the interplay between our self-definition and the constantly changing environment.

The above discussion raised the question of whether the environment has the power to provide consolation or signify an escape route. The analysis of both authors proves that not only is the genre of new nature writing more than suitable for such exploration but that the environment, if understood as a co-creation or an interface of the human and the natural, significantly enriches our identity. Moreover, the validity of the author's experience makes both environmental texts convincing. Deakin's swimming journey enriches his personhood with a new type of experiential knowledge; Thomson's walk through the English countryside helps him, among other things, understand the volatile dimension of his identity. The basis of the consolatory role of nature is in meeting "the overwhelming need [...] for the envisioning of the framework which would allow positive manifestations of forms of expression in ways which are harmonious with the needs of the natural world" (Kidner 72). Therefore, the ability to identify as part of the environment and to understand its rich cultural symbolism roots our existence in an entity that is fluid and flexible.

Bibliography

Abberley, Will, et al. *Modern British Nature Writing, 1789–2020: Land Lines.* Cambridge University Press, 2022.

Alexander, Neal. "Theologies of the Wild: Contemporary Landscape Writing." *Journal of Modern Literature,* vol. 38, no. 4, 2015, pp. 1–19.

Attlee, James. "The Green Road into the Trees: An Exploration of England, By Hugh Thomson." *The Independent,* 29 June 2012. https://www.independent.co.uk/arts

-entertainment/books/reviews/the-green-road-into-the-trees-an-exploration-of
-england-by-hugh-thomson-7899514.html. Accessed 4 Nov. 2023.

Clark, Timothy. *The Cambridge Introduction to Literature and Environment.* Cambridge University Press, 2011.

Deakin, Roger. *Waterlog.* 1999. Vintage Books, 2014.

Dudley, Marianna. "Reflections on Water: Knowing a River." *RCC Perspectives No. 4. Environmental Knowledge, Environmental Politics: Case Studies from Canada and Western Europe,* edited by Jonathan Clapperton and Liza Piper, 2016, pp. 47–54. https://www.environmentandsociety.org/sites/default/files/2015_i4_final-07_dudley_0.pdf.

Hampton, Alexander J. B. "Post-secular Nature and the New Nature Writing." *Christianity and Literature,* vol. 67, no. 3, 2018, pp. 454–471.

Kidner, David W. "Culture and the Unconscious in the Environmental Ethics." *Environmental Ethics,* vol. 20, no. 1, 1998, pp. 61–80.

Lorimer, Jamie. *Wildlife in the Anthropocene.* University of Minnesota Press, 2015.

Mabey, Richard, editor. *The Oxford Book of Nature Writing.* Oxford University Press, 1997.

Mabey, Richard. *Nature Cure.* 2005. Vintage Books, 2015.

Macfarlane, Robert. *The Old Ways: A Journey on Foot.* Hamish Hamilton, 2012.

Smith, Jos. *The New Nature Writing: Rethinking the Literature of Place.* Bloomsbury Academic, 2017.

Thomson, Hugh. *The Green Road into the Trees: A Walk through England.* 2012. Windmill Books, 2013.

Williams, Raymond. *Keywords.* Flamingo, 1993.

CHAPTER 8

"Making Things Right": Fostering Relationships of Care in Narratives of Feminist Hydrocommons

Aleksandra Kamińska

Abstract

"Water," as Greta Gaard (2017) points out, "is a feminist issue" (81). This diagnosis not only reflects the popular trope of linking women and water which is present in many global cultures (see Staniland 2023), but also surfaces as water's gendered materiality influences the intersections of environmental systems and biological bodies. Astrida Neimanis in her *Bodies of Water: Posthuman Feminist Phenomenology* (2017) stresses water's capacity to connect female bodies "to other bodies, to other worlds beyond our human selves," thus presenting an opportunity to challenge anthropocentrism by exploring our involvement in "a more-than-human hydrocommons" (2). In this perspective, the link between female bodies and water becomes a sphere of potential connectivity and agency, offering an opportunity to explore postmothering relationships of care (see Phillips 2016). The chapter focuses on two literary explorations of feminist hydrocommons: Robin Wall Kimmerer's non-fiction work *Braiding Sweetgrass* (2013) and Joanna Bator's short fiction "Tikkun Olam" (2022), arguing that both texts present the potential for renewal, reclaiming agency and reconnecting with other (both human and other-than-human) bodies through the narratives of women's custody of bodies of water. It is argued that in both works in question, the authors focus on the sense of purpose and consolation stemming from the refiguration of multivarious entanglements as networks of mutual relationships rooted in care, which in turn allows them to transgress "motherhood environmentalism" frequently ascribed to feminist narratives and instead explore the liberating potential of experiencing embodied more/than/human relatedness.

Keywords

water – material feminism – hydrocommons – Robin Wall Kimmerer – Joanna Bator

1   Introduction: Finding Consolation through Care

In her 2019 Nobel Lecture, Olga Tokarczuk observed that our current reality is dominated by first-person narratives, which focus on the autonomy of the individual at the cost of "building an opposition between the self and the world" (Tokarczuk). To combat this tendency and find protection against the resulting feeling of alienation, she advocates narrative modes rooted in tenderness. For the author of *The Books of Jacob*, tenderness "is a way of looking that shows the world as being alive, living, interconnected, cooperating with, and codependent on itself" (Tokarczuk); it is a mode that can potentially shield us from the modern condition, flooded with "anxiety oozing from all directions" (Tokarczuk). Consequently, Tokarczuk stresses the consolatory potential of narratives of interconnectedness that foreground codependencies, encouraging readers to find solace in the ways our lives are enmeshed with those of others, including both humans and the more-than-human world. It is my contention that the two texts discussed in this chapter offer an interesting perspective on how this consoling effect can be achieved also in first-person narratives that focus on expanding the sense of self through the experience of physical immersion and transcorporeality. In Robin Wall Kimmerer's *Braiding Sweetgrass* and Joanna Bator's short story "Tikkun Olam," female narrators/protagonists realise their participatory presence in the living mesh of entangled lifeforms through establishing and fostering close relationships with bodies of water. By learning to care for the respective ponds with which they interact, they redefine their perception of the surrounding reality and rediscover a sense of belonging in the world, rooted in the attitude of responsibility towards other watery beings; moreover, they experience a profound feeling of consolation stemming from the realisation of shared watery embodiment, making the care reciprocal.

2   Watery (Re)Connections

In her *Bodies of Water: Posthuman Feminist Phenomenology* (2017), Astrida Neimanis proposes a reflection on watery embodiment as a way of discovering connection to other bodies and other worlds, transgressing the boundaries of discrete individuals and species. As she points out, understanding and appreciating watery connectivity compels us to acknowledge that our corporeal borders are never fixed, as the body of water is "multiscalar and multigenerational, porous and palimpsestic" (29). Starting from the realisation that all beings are embodied water in the biological, physical sense, Neimanis argues that shared watery embodiment allows us to experience our connection body to body, not only across time, but also across the division between the human and the other-than-human.

Neimanis draws on the watery materiality of gestation, explaining that biological processes of life proliferation are "maintained by [...] water relays" (Neimanis, *Bodies of Water* 66). Amniotic waters are here envisaged as part of the flux of water that enables the creation of new life but also immediately connects it with the "biological and meteorological cycles" (*Bodies of Water* 66), as water sustains both the foetus and the gestating mother. The awareness of this process of embeddedness of gestation within the larger body of planetary waters can, according to Neimanis, make us "more thoughtful, and more responsive, in terms of what we give back to water in all its forms, but in particular to those planetary water bodies that we [...] currently exploit, pollute, and instrumentalize" (*Bodies of Water* 67). In other words, this watery view of gestation is configured as a means of transferring from the mode of exploitation towards the attitude of caring stewardship rooted in the acknowledgement of shared embodiment of both human and other-than-human subjects. It draws on Stacy Alaimo's argument that "thinking across bodies" compels us to readjust our perception of the world and shift from the anthropocentric perspective towards the realisation that, instead of constituting a resource to be explored by humans, the environment is, "in fact, a world of fleshy beings, with their own needs, claims, and actions" (Alaimo, "Trans-corporeal Feminisms" 238), in which bodies constitute both agents and objects of environmental change (Alaimo, *Bodily Natures* 89–94). Consequently, Alaimo insists, acknowledging our shared transcorporeality (defined as "the time-space where human corporeality, in all its material fleshiness, is inseparable from 'nature' or 'environment'" ["Trans-corporeal Feminisms" 238]) opens up "crucial ethical and political possibilities" ("Trans-corporeal Feminisms" 238).

In her book, Neimanis frames this ethical perspective as viewing the human as a part of the hydrocommons – a shared community based on the ever-present watery exchange between different bodies and the realisation that, in the cycle of life, bodies are always indebted to other bodies:

> We are created in water, we gestate in water, we are born into the atmosphere of the same water although more diffuse, we take in water, we harbour it, it sustains and protects us, it leaves us [...] at the same time as we are always, to some extent, in it. The passage from body of water to body of water (always *as* body of water) is never synecdochal or metaphoric; it is radically material. These complex and shared cyclings – body, to body, to body – comprise our planetary hydrocommons. (Neimanis, *Bodies of Water* 86)

Therefore, in this water-based perspective, bodies appear as much less fixed and clearly separated from one another; instead, Neimanis encourages thinking in terms of sharing and exchanging – across time, across space, and across species.

Finally, for Neimanis – although the planetary hydrocommons encompasses each and every body – the shared watery embodiment is a predominantly feminist perspective. In declaring that a "material feminine womb [...] reverberates of and with the unknowability of planetary waters" (*Bodies of Water* 84), she explores the connection between the feminine element and the notion of watery gestation perceived in the larger, posthuman frame. In this way, Neimanis' theory transcends the traditional feminist association of the female body with the body of water which, as Elizabeth Stephens (2014) points out, concentrates on women's bodily fluids (tears, menstrual blood, breast milk, female ejaculate, amniotic fluids, etc. [Stephens 186]) and is aptly reduced by Emma Staniland to the simple equation "woman = water = womb = life" (Staniland 5). Instead, Neimanis argues in favour of what she calls "posthuman gestationality," defined as "a facilitative mode of being, but one that is not necessarily tied to the female body" (Neimanis, *Bodies of Water* 68–69). In fact, as she explains, "water is [...] the gestational element" (Neimanis, "Hydrofeminism" 87) as "[g]estational waters [...] participate in the greater element of planetary water that continues to sustain us, protect us, and nurture us, both extra- and intercorporeally [...] Water connects the human scale to other scales of life" (Neimanis, "Hydrofeminism" 87). Such a view of gestationality opens up the space for ethics of motherhood extending beyond the biology of birthing, focusing instead on an ethics of caring as well as the reciprocity in human–other-than-human relations.

Arguably, Neimanis' concept of the hydrocommons offers an interesting background for investigating how the relationship between women and other bodies of water is framed in literary narratives. Two such narratives will be examined in the following part of the chapter, juxtaposing a work of nonfiction with an example of short literary fiction. Although published within different literary genres and emerging from different cultures, these two narratives share interesting parallels in terms of narrating their female protagonists' involvement in the hydrocommons.

3   Making a Good Home Together: *Braiding Sweetgrass*

An interesting exploration of the hydrocommons in non-fiction can be found in *Braiding Sweetgrass: Indigenous Wisdom, Scientific Knowledge and the*

Teachings of Plants (2013) by Robin Wall Kimmerer, an American scientist and author of Potawatomi descent. As a researcher, Kimmerer specialises in moss biology; she is also a Native American activist and advocate for bringing Native Peoples' wisdom into environmental discourse. In her environmental writings, she adopts multiple perspectives as she "shares stories of encounters with plants as a mother, scientist, botanist, tribal citizen, and teacher" (Aalto 146) in order to encourage a "relationship of reciprocity" (Aalto 147) between human and other-than-human participants in the natural world. In *Braiding Sweetgrass*, Kimmerer combines elements of autobiography, family history, botany, and Native American folklore as she braids together a narrative with a strong environmental message.

An important theme in Kimmerer's work is her preoccupation with language and her awareness that "in wild places, we are audience to conversations in a language not our own" (Kimmerer, *Braiding Sweetgrass* 48; *Democracy of Species* 2), combined with an intense "longing to comprehend this language" (*Braiding Sweetgrass* 48; *Democracy of Species* 2). As she explains in the chapter titled "Learning the Grammar of Animacy," more recently reprinted in the small volume *The Democracy of Species* (2021), relying solely on the language of science in our understanding of nature – despite "the richness of its vocabulary and its descriptive power" (*Braiding Sweetgrass* 49; *Democracy of Species* 2) – must lead to "a grave loss in translation" (*Braiding Sweetgrass* 49; *Democracy of Species* 3). The author manages to progress towards a more natural language through exploring her Native American heritage and painstakingly acquiring Potawatomi, which she never learnt as a child. In this new language, she discovers the titular "grammar of animacy": here, the word "water" is a verb rather than a noun, which stresses its indefinite, animate, processual character: the fact that "for this moment, the living water has decided to shelter itself" in this particular place, but it could also "do otherwise" (*Braiding Sweetgrass* 55). Through transforming "the metaphors we live by," the writer enters into the realm of language which fosters the awareness of the ever-changing nature of our bodies of water, engaged in the "constant process of intake, transformation, and exchange" (Neimanis, *Bodies of Water* 2).

But the greatest experience of connectivity between the human and the other-than-human in Kimmerer's book is presented through the image of the writer's physical interaction with a specific body of water. In the chapter titled "A Mother's Work" Kimmerer relates her experience of recultivating a pond on her property as she was trying to set up a new home for herself and her daughters after her relationship with her partner fell through. From the outset, Kimmerer establishes a parallel between the pond and herself as she relates how she associated the stewardship of "what was described as a trout pond" in the real estate advertisement with getting her life into order again ("This was the one place where I somehow felt as if I could make things right" [*Braiding Sweetgrass* 88]) and her ambitions as a (single) mother: "being the good mother, good enough for two parents, seemed within my grasp" – all that she needed to achieve in order to attain this goal was a "swimmable pond" (*Braiding Sweetgrass* 83) for her daughters to enjoy.[1] Yet, as she soon found out, the pond was far from swimmable: in fact, it was overgrown with algae to the point where "you could not tell where weeds left off and water began" (*Braiding Sweetgrass* 83).

As a biologist, Kimmerer first approached the task by relying on her scientific and professional knowledge: she started by researching the topic of pond rehabilitation and identifying the exact species of algae she was dealing with by testing samples under her microscope. Interestingly, even at this early stage, she perceived the algae as potential allies in her recultivation project:

> In this single tuft were long threads of *Cladophora*, shining like satin ribbon. Wound around them were translucent strands of *Spirogyra*, in which the chloroplasts spiral like a green staircase. The whole green field was in motion, with iridescent tumbleweeds of *Volvox* and pulsing euglenoids stretching their way among the strands. *So much life in a single drop of water* [...] Here were my *partners in restoration*. (*Braiding Sweetgrass* 88, my emphasis)

In this strikingly visual image, Kimmerer showcases the living, animate character of the algae, which are shown "in motion," "pulsing" and "stretching," engaged in their own life processes and pursuing their independent goals. While realising this, Kimmerer already begins to perceive their shared situation as that of "a more-than-human hydrocommons" (Neimanis, *Bodies of Water* 2), in which her own body and the bodies of the algae are connected; she also imagines the possibility that their agendas could potentially overlap.

Yet, despite this cooperative attitude, Kimmerer soon realised the sacrifices she would have to make in order to see her project through: as she was attempting to recultivate the pond and make it swimmable for herself and her children, she was destroying the lives and habitats of many smaller organisms,

---

1 This internalised sense of obligation potentially hints at the more problematic association between women and water indicated by Gaard (2017): the gendered role of supplying their families with water takes a toll on many women across the globe, not only adding to their daily workload, but also negatively affecting their health and educational opportunities as well as exposing them to gender-based violence (see Concern Worldwide 2022).

which she found hard to accept. Interestingly, she only managed to overcome these difficulties through physical immersion in the pond, thus blurring the boundaries between different bodies of water. While initially she walks into the water dressed in protective garments, the physical barriers between herself and the pond begin to shift almost immediately: "The rubber boots that were intended to keep the pond at bay now contain it. And me. And one tadpole," she reports (*Braiding Sweetgrass* 82). Soon she is ready to become fully immersed and discovers the emancipating potential of the experience: "I remember the liberation of just walking right in to my waist the first time, the lightness of my T-shirt floating around me, the swirl of water against my bare skin. I finally felt at home" (*Braiding Sweetgrass* 89).

From that moment onwards, Kimmerer experiences her connection with the pond in a visceral, physical sense. Through reconnecting with the watery environment, she establishes a durable and reciprocal relationship. On the one hand, she cares for the wellbeing of the pond, but she also experiences its care:

> The pond built my muscles, wove my baskets, mulched my garden, made my tea [...] Our lives became entwined in ways both material and spiritual. It's been a balanced exchange: I worked on the pond and the pond worked on me, and together we made a good home. (*Braiding Sweetgrass* 95)

In the passage above, through the use of the active voice in relating how the pond "builds," "weaves," "mulches" and "makes" things, Kimmerer's narrative construes the pond as an active subject, on a par with the narrator herself. The "good home" is expanded to include not only the dwelling of the human inhabitants, but also the habitat of the other-than-human bodies which live there. As Neimanis explains, a more-than-human hydrocommons challenges the anthropocentric perspective; in acknowledging the other-than-human body of water as an active participant in shaping the shared home, Kimmerer embraces this broader outlook.

In positioning herself as the carer for the pond, Kimmerer draws on her Native American heritage by evoking the Potawatomi tradition in which women are regarded as the "Keepers of Water," whose responsibility is to carry sacred water to ceremonies, but also to safeguard water for their family members and other-than-human animals under their care (Kimmerer, *Braiding Sweetgrass* 94). At the same time, by foregrounding the agency of the pond as well as the algae that live in it, Kimmerer evokes the indigenous perspective in which "all living things contain spirit" and thus are "active members of society," exercising their ability to "interpret, understand and implement" (Watts 23; see also Todd, "Indigenizing the Anthropocene," Chandler and Reid, *Becoming Indigenous*). By stressing how she receives care from the pond in equal measure as she herself cares for that body of water, Kimmerer relies on the indigenous tradition (rather than on Euro-Western academic posthumanist and speculative discourses) to take a non-anthropocentric stance.

Yet out of the three ways in which the author lives her watery involvement (through her use of science, rediscovering her ties to her Native American heritage, and physical action of caring for the pond), it is this third type of experience that most significantly shapes her perspective. The physical engagement with water expands Kimmerer's experience of connection with other living beings and, consequently, her sense of responsibility and care:

> The circle of care grows larger and caregiving for my little pond spills over to caregiving for other waters. The outlet from my pond runs downhill to my good neighbour's pond. What I do here matters. Everybody lives downstream. My pond drains to the brook, to the creek, to a great and needful lake. The water net connects us all. (*Braiding Sweetgrass* 97)

In her final realisation, the narrator embraces the "hydrocommons of wet relations" (Neimanis, *Bodies of Water* 4), acknowledging the ultimate connectivity of the bodies of water and extending the "water net" to not only include the bodies in immediate proximity, but also those further away on the grounds of our shared embodiment. At the same time, Kimmerer associates her sense of ethical responsibility towards other bodies with expanding the limits of motherhood and motherly care. While initially her main concern was "fixing the pond for [her] kids" (*Braiding Sweetgrass* 86), and in this endeavour she was clearly prioritising the needs of her own daughters, her agenda has eventually changed. Having experienced the watery reconnection to the surrounding living environment, she is ready to blur the lines between human and other-than-human children and wants to nurture also "frog children, nestlings, goslings, seedlings, and spores" (*Braiding Sweetgrass* 97). Her experience of motherhood extends "beyond her family, beyond the human community, embracing the planet, mothering the earth" (*Braiding Sweetgrass* 97), thus paralleling Neimanis' postulate of posthuman gestationality – one that is no longer "tied to the female human," but which focuses on what we "give back to water in all its forms" (Neimanis, *Bodies of Water* 69).

4   Renewal through Water: "Tikkun Olam"

A good example of a work of fiction in which the theme of hydrocommons is explored through a narrative of a woman's stewardship of a body of water is the short story "Tikkun Olam" by Joanna Bator, an award-winning Polish writer. Bator's works have been translated into several languages; in 2018, together with her German translator Esther Kinsky, she was awarded the Calw Hermann Hesse Prize. "Tikkun Olam" is the final story included in the volume *Ucieczka niedźwiedzicy* (2022).

It can be argued that the watery reconnection of the main character in "Tikkun Olam" takes on a consolatory turn, exploring one of the traditional meanings of water as the symbol of "the cycle of birth, life, death and return" (Kattau 122). The unnamed protagonist and narrator is a post-menopausal woman, who is physically active and enjoys being outdoors and swimming. In the opening scene she is emerging from a pond and regaining consciousness: "Przebudziłam się na brzegu, ale od pasa w dół wciąż spoczywałam w wodzie" [I came to on the shore but the lower half of my body was still submerged in water][2] (Bator 301). She is enthusiastically welcomed back to the world by her elderly dog Bułka. Then the narrative rewinds and more details are divulged about the pond and the land surrounding it.

The narrator has bought the land that used to belong to a now-closed brick factory, with three clay pit ponds, created after abandoned pits where clay had been mined filled with rain and groundwater, forming new watery habitats. Not unlike Kimmerer in her account of pond recultivation, the narrator in "Tikkun Olam" assumes responsibility for the ponds and makes an effort to bring them back to life:

> Potem przez wiele lat wody zarastały zielskiem i służyły jako niele-galne śmietnisko, wędkarze coś w nich łowili, dzieci kąpały się i topiły. Zwierzętom w topieniu się pomagali ludzie. Dno zbiorników kryło wiele dziwnych rzeczy, które wydobył przemysłowy magnes i dwóch wynajętych nurków z odpowiednim sprzętem. Wyciągnięto metalowy wagonik na węgiel [...] stare rowery i zlewy, sedesy i kuchenki, maszyny do szycia, potłuczone kafle i zardzewiałe krany, mnóstwo butelek i puszek po piwie. Były też worki pełne zwierzęcych kości [...].
>
> To, co martwe, pochowałam w ziemi na brzegu, *a to, co na Gliniankach zastałam żywe, starałam się ochronić.* Nowe drzewa łatwo się zakorzeniły w wilgotnej glebie, w ciągu kilku lat stając się prawie lasem. *Woda oczysz-czona ze śmieci i okrucieństwa odżyła,* ryby mnożyły się [...]. (Bator 303, my emphasis)

---

2   All quotations from Joanna Bator's work are in my translation (A. K.).

[Then, for many years, the ponds were overgrown by weeds and used as illegal garbage dumping sites; anglers would come here to fish, children would swim and drown here. Animals were drowned, assisted by people. The bottom of the ponds was strewn with strange objects that were extracted with the help of an industrial magnet and two hired divers with professional equipment. They removed a coal minecart from a Silesian mine [...] old bikes and kitchen sinks, toilet bowls and sewing machines, broken tiles and rusty taps, heaps of empty beer cans and bottles. There were also bags with animal bones [...].
    What was dead I buried on the shore, and *what I found alive in Glinia-nki I tried to protect.* New trees easily put down roots in the damp soil, growing almost into a forest over the last few years. *Purified from garbage and cruelty, water became alive again*; fish multiplied [...]. (Bator 303, my emphasis)]

Further in the story, the narrator explains how she became the custodian of this plot of land and developed a relationship with it. She describes how it has changed under her care. She takes time to enumerate different species that live there: ducks, coots and swans, two crested grebes, a spotted eagle, a marsh harrier, foxes and a family of beavers. A pair of teacup pigs are purposely left by someone on the property; they are adopted by the narrator's dog Bułka and together with the narrator they appear to form a more-than-human found family unit.
    The ways in which the narrator takes care of the ponds and the surrounding land are based on co-existence and cooperation. She uses willow twigs from a willow tree overthrown by beavers to form a fence and the twigs soon sprout back to life; she also takes effort to leave a small gap under the fence to allow passage for small animals coming from the fields, who are also settling in around the ponds. Every morning, together with her dog, the narrator does her rounds, taking care of her land and its inhabitants. They investigate every nook and cranny, looking under the trees and into the bushes, traversing the ponds and keeping stock of every change: every new animal, a new mushroom growing on a tree trunk, new developments on the beavers' dam, leftovers of someone's night hunt. But her most meaningful, intimate contact with the sur-roundings is established through swimming in the pond.
    The ending of the short story is construed in a dream-like narrative: the narrator goes night swimming during a summer tempest, is swallowed by a gigantic catfish living in one of the ponds, and then spat out on the shore where she returns to life, transformed by the experience. Her symbolic rebirth,

on the one hand, alludes to the mythological power of water as an agent of
spiritual cleansing and renewal (see e.g. Witzel 20–21), evoked, among others,
in the Christian ritual of baptism; on the other hand, it takes on a very physical
dimension and occurs through strong interconnection with nature:

> [w] *uszy wpływały* szelesty i pluski Glinianek. [...] Udało mi się wyczołgać
> na brzeg i zwinąć jak embrion w cieniu wierzby. Trawa była wilgotna, ciepła,
> miękka, a pośród niej toczyło się intensywne małe życie, *przenikające we*
> *mnie.* (Bator 314–315)

> [the murmurs and splashings of Glinianki *poured into my ears* [...] I man-
> aged to crawl out of the water and curled up like an embryo in the shade
> of the willow. The grass was damp, warm and soft, and among it an intense
> little life throbbed, *seeping into me.* (Bator 314–315, my emphasis)]

After reemerging from the water, the narrator experiences a transformation of
perception linked to a heightened sense of interconnection and transcorporeality; her body and consciousness are blurring the boundaries between herself
and the living world that surrounds her:

> Byłam jak puste naczynie, ciało dopiero co stworzone z wody, gliny,
> ognia i powietrza, ze ściankami *wciąż przepuszczalnymi jak membrana.*
> [...] Świat *przepływał koło mnie, przeze mnie,* byłam zwierzęciem,
> człowiekiem, wszystkimi żywiołami [...] (Bator 315)

> [I was like an empty vessel, a body freshly made from water, clay, fire and
> air, its shell *still permeable like a membrane.* [...] The world *flowed around*
> *me and through me,* I was an animal, I was a human, I was the elements [...].
> (Bator 315, my emphasis)]

At the end of the story, the narrator experiences an epiphany. Taking care
of living things and blurring the boundary between human and other-than-
human, she is ready to return to society and resume her responsibilities after
a temporary hiatus necessitated by professional burnout; she is already imagining herself undertaking creative work. The title of the short story, "Tikkun
Olam," is a Hebrew term signifying an action aimed at improving the world.
In Bator's text, the way to undertake this momentous task is through a watery
reconnection, experiencing and acknowledging one's porousness, and taking
one's place in the hydrocommons.

5   Conclusion: Watery Narratives of Care

To conclude, it can be argued that these two texts represent ways of finding consolation through actively participating in hydrocommons. For both Robin Wall Kimmerer and the nameless narrator in Bator's short story, their ownership of the ponds becomes a way to heal from, respectively, the trauma and heartbreak of a broken relationship and family, and work burnout. At the beginning of their narratives both protagonists have experienced loss and find themselves at a crossroads, hoping to let go of the past. For both of them the sense of belonging discovered through the affiliation with the hydrocommons becomes a source of profound consolation, achieved chiefly through bodily immersion. In both narratives, the physical engagement in taking ownership *and* stewardship of the land is an important element, but the main transformative moment in both texts occurs when the protagonists enter their ponds, immersing themselves in the water and connecting with the other-than-human element: Kimmerer becomes ready to extend her experience of motherhood to include other-than-human babies; Bator's protagonist – in a more metaphorical way – for a moment *becomes* the other-than-human as she is symbolically reborn from the pond for which she has cared. The permeable membranes of the narrators' bodies allow movement in both directions: as their experience of self is expanding, so, too, they become ready to acknowledge the seeping of the other into themselves. Both texts, in fact, make use of mythological elements (the Native American story of the Keeper of Water; the Japanese myth of the giant catfish), combining them with practical, environmental descriptions of recultivating neglected or polluted watery habitats.

In addition, through these acts of connecting – to the ponds, and, by extension, to the environment – both narrators experience re-connection to their own lives. Kimmerer makes peace with the fact that her daughters are now leaving home and herself to go their independent ways, and finds herself ready to enter the next chapter of life, extending her maternal care to the land and its other-than-human inhabitants, while Bator's narrator feels ready to overcome her professional crisis and re-engage with the world, already imagining the new buildings she is going to design and renovate (she is an architect).

At the same time, through examining the relationship of individual protagonists with specific bodies of water, the two authors engage with what Neimanis terms "watered politics of location," centred on scrutinising the way in which we are "specifically situated in relation to specific waters" ("Feminist Subjectivity" 36). Depicting the caring, reciprocal relationships of the two women with the ponds whose guardians they become, both texts offer a shift in narrative from positioning women predominantly as victims of gender inequalities involved in the ways in which water is not only distributed and controlled as a vital resource, but also perceived and reflected in cultural representations (see Lahiri-Dutt), and instead showcase female protagonists becoming valuable participants in the hydrocommons, both willing and able to take the step from "thinking *about*" water and moving towards "thinking *with*" water (see Chen 276) – from treating the ponds as a resource to be explored, or commodity to be owned, towards a relationship of mutual support.

Both protagonists move beyond merely recognising their watery embodiment as a biological fact and actively embrace "living their embodiment as watery" (Neimanis, "Feminist Subjectivity" 24), which can be perceived as an overt confirmation of their allegiance to the hydrocommons. As Chen, MacLeod and Neimanis explain, the way we are situating water – and situating ourselves *in relation to* water – "may serve to acknowledge or deny our participation" ("Introduction" 9); consequently, the willingness of the two protagonists to embrace responsibility for their respective ponds becomes not only a personal, but also an ethical and political gesture.

Finally, it is interesting to note that both protagonists are middle-aged, possibly post-menopausal women, which automatically situates them outside the typical "woman = water = womb = life" paradigm pointed out by Staniland. In this way, both these narratives – while touching on themes of birthing and mothering – evade the pattern of what has been criticised as "motherhood environmentalism" (see Phillips 474), where care and nurturing are framed as "womanly" values and women's relationship towards the natural world is imagined as that of self-sacrificing mothers. Instead, both Kimmerer and Bator offer images of "posthuman gestation" which "connects the human scale to other scales of life" (Neimanis, "Feminist Subjectivity" 87), a reciprocal relationship rooted in caring conveyed as a "recognition of our entangled materialities" (Phillips 471) and care practices which "maintain, promote or enhance the flourishing of relevant parties" (Phillips 476). Consequently, both texts can be viewed as exploring the posthuman, postmaternal gestation, "reconfigured through radical and liberatory focus on embodied relatedness" (Phillips 468), a relationship forged and sustained not so much through a biological mechanism of gestation involving specific bodily fluids, as through the experience of the physical embeddedness of the protagonists' bodies of water in the hydrocommons.

Bibliography

Aalto, Kathryn. *Writing Wild: Women Poets, Ramblers, and Mavericks Who Shape How We See the Natural World.* Timber Press, 2020.

Alaimo, Stacy. "Trans-corporeal Feminisms and the Ethical Space of Nature." *Material Feminisms*, edited by Stacy Alaimo and Susan Hekman, Indiana University Press, 2008, pp. 237–264.

Alaimo, Stacy. *Bodily Natures: Science, Environment, and the Material Self.* Indiana University Press, 2010.

Bator, Joanna. "Tikkun Olam." *Ucieczka niedźwiedzicy*. Znak, 2022, pp. 301–316.

Chandler, David, and Julian Reid. *Becoming Indigenous: Governing Imaginaries in the Anthropocene*. Rowman & Littlefield, 2019.

Chen, Cecilia. "Mapping Waters: Thinking with Watery Places." *Thinking with Water*, edited by Cecilia Chen, Janine Macleod and Astrida Neimanis, McGill-Queen's University Press, 2013, pp. 274–298.

Chen, Cecilia, Janine MacLeod, and Astrida Neimanis. "Introduction: Toward a Hydrological Turn?" *Thinking with Water*, edited by Cecilia Chen, Janine MacLeod and Astrida Neimanis, McGill-Queen's University Press, 2013, pp. 3–22.

Concern Worldwide. *5 Reasons Why Water is a Women's Issue*, 7 March 2022. n. p. https://www.concern.net/news/water-is-a-womens-issue. Accessed 15 February 2025.

Gaard, Greta. "Feminism and Environmental Justice." *The Routledge Handbook of Environmental Justice*, edited by Ryan Holifield, Jayajit Chakraborty and Gordon Walker, Routledge, 2017, pp. 74–88.

Kattau, Colleen. "Women, Water and the Reclamation of the Feminine." *Wagadu: A Journal of Transnational Women's and Gender Studies*, vol. 3, no. 1, 2006, pp. 114–143.

Kimmerer, Robin Wall. *Braiding Sweetgrass: Indigenous Wisdom, Scientific Knowledge and the Teachings of Plants*. Penguin Books, 2013.

Kimmerer, Robin Wall. *The Democracy of Species*. Penguin Random House, 2021.

Lahiri-Dutt, Kuntala, editor. *Fluid Bonds: Views on Gender and Water*. Bhatkal and Sen, 2006.

Neimanis, Astrida. "Hydrofeminism: Or, On Becoming a Body of Water." *Undutiful Daughters: New Directions in Feminist Thought and Practice*, edited by Henriette Gunkel, Chrysanthi Nigianni and Fanny Söderbäck, Palgrave Macmillan, 2012, pp. 85–100.

Neimanis, Astrida. "Feminist Subjectivity, Watered." *Feminist Review*, no. 103, 2013, pp. 23–41.

Neimanis, Astrida. *Bodies of Water: Posthuman Feminist Phenomenology*. Bloomsbury, 2017.

Phillips, Mary. "Embodied Care and Planet Earth: Ecofeminism, Maternalism and Post-maternalism." *Australian Feminist Studies*, vol. 31, no. 90, 2016, pp. 468–485.

Staniland, Emma, editor. *Women and Water in Global Fiction*. Routledge, 2023.

Stephens, Elizabeth. "Feminism and New Materialism: The Matter of Fluidity." *InterAlia*, vol. 9, 2014, pp. 186–202.

Todd, Zoe. "Indigenizing the Anthropocene." *Art in the Anthropocene: Encounters Among Aesthetics, Politics, Environment and Epistemology*, edited by Heather Davis and Etienne Turpin, Open Humanities Press, 2015, pp. 241–254.

Tokarczuk, Olga. *Nobel Lecture: The Tender Narrator.* 2019, n. p. https://www.nobelprize .org/prizes/literature/2018/tokarczuk/lecture/. Accessed 9 January 2024.

Watts, Vanessa. "Indigenous Place-Thought and Agency amongst Humans and Non-Humans (First Woman and Sky Woman Go on a European World Tour!)." *Decolonization: Indigeneity, Education and Society,* vol. 2, no. 1, 2013, pp. 20–34.

Witzel, Michael. "Water in Mythology." *Daedalus,* vol. 144, no. 3, 2015, pp. 18–26.

PART 3

Cultivating Nature as a Way of Fostering Communal Justice and Redress

∵

Gardening Forking Paths: the Figure of the Rambunctious Garden and Relational Ethics of Attentiveness and Care

Anton Belenetskyi

Abstract

This chapter aims to study how and to what effect the ages-old notion of nature is being imaginatively redefined by selected works of the 21st-century U.S. literature in response to the complex uncertainty elicited by the advent of the new geological epoch of the Anthropocene, brought on by the rapidly unfolding anthropogenic ecological breakdown. In so doing, it regards such uncertainty as an opening of ethical possibility and examines what better ways of relating and relating to the world beyond the human may emerge out of the newly volatile idea of nature set into motion by the ongoing reckoning with what Lynn Keller defines as the "self-conscious Anthropocene," at once a vertiginous realisation and a radical reorganisation of the planetary interrelations. Drawing on Emma Marris's theorisation of the epoch's uncertain natures as "rambunctious gardens," the ongoing redefinition of the human–nature interrelatedness is situated in the inherently fluctuating figure of the garden, simultaneously natural and cultural, and the practices of attentiveness and care that are commonly associated both with the process of gardening and the more-than-human relational ethics. Weaving together four highly variegated texts centred around gardens and gardening (David Searcy's 2001 novel *Ordinary Horror*, Robin Wall Kimmerer's 2013 treatise *Braiding Sweetgrass*, Mei-mei Berssenbrugge's 2013 poetry collection *Hello, the Roses*, and Camille T. Dungy's 2023 memoir *Soil*), the chapter elaborates on how attending to the elusive reconfigurations of the concept of nature in the Anthropocene may help us become reconciled to the world's irresolvable uncertainty and develop an ethic of tending its contingently forking paths towards a better future of coexistence and co-creation.

Keywords

the Anthropocene – attentiveness – care – garden – relational ethics – uncertainty

...

We have, finally, no clear or distinct ideas.

DONNA J. HARAWAY, "Situated Knowledges"

..
.

1   Introduction: Whither Now for Nature?

Nature has always been a notoriously thorny notion. Raymond Williams even famously quipped that it "is perhaps the most complex word in the [English] language" ("Nature" 164): an infinitely troublesome concept with equally end-less possibilities for interpretation. For some, it is antiquated beyond repair and reeks of essentialism; for others, it brims with a world-making potential and offers a source of hope when hope is otherwise scarce. For Williams, how-ever, the irreducible value of nature as a conceptualisation through which the reality beyond the human is approached, understood, and potentially acted upon lies precisely in this complicated plurality that translates into its potent ambivalence. What happens when we finally become, inadvertently following Williams's admonition, "especially aware of its difficulty" ("Nature" 169)? This chapter grows out of the interest in what happens when the still-widespread commonsensical notion of nature – the uncritical vision thereof as that which is "opposed to humans or human creations" ("nature," *Oxford English Diction-ary*) – is confronted with the complexities of our present-day reality of the multifaceted ecological breakdown where the natural and the human worlds appear to coalesce in the contingent choreography of co-creation, sometimes quite ruinous, at other times tentatively promising.

For nature, both as an idea and a material realm, has arguably never been more uncertain than it is now: at once familiar and strange, comforting and overwhelming, enduring and endangered. Nature's ambiguity, which Williams championed in the far-off year of 1976, has ever since spilt far beyond the aca-demic debate and is now an acute part of our day-to-day experience. Its bor-ders are unprecedentedly porous, its definitions more and more contentious, its forking futures ever harder to predict and even harder to remain hopeful about. When the effects of climate change, dramatic biodiversity loss, and too many other, pronouncedly unnatural cataclysms are unabatedly unfold-ing before our very eyes, of what use may the elusive idea of nature be? What solace can it promise?

Following Williams's seminal analysis, my attentiveness to the ongoing trials and tribulations of now-volatile nature is rooted in the belief that the deepening uncertainty that surrounds the notion is what enables it to remain relevant and intellectually generative in today's fast-deteriorating world. On the one hand, such uncertainty, thanks to its destabilising powers, may reveal the notion's heretofore rarely questioned internal contradictions. On the other hand, by unsettling nature's assumed human/environment dichotomy, it brings to the forefront the sheer dynamic variability of the notion's messy poetics and politics that may serve as a surprisingly fruitful framework for changing our taken-for-granted approaches to the other-than-human. Either way, the present-day transformations of the newly uncertain concept of nature, no longer intrinsically self-evident yet far from being written off as conceptually obsolete, merit our utmost attention as they may offer us a glimpse into new possible ways of making sense of both the more-than-human world and the striking novelty of its radical interrelatedness. As Williams aptly puts it: "We need different ideas [of nature] because we need different relationships" ("Ideas of Nature" 85). The chapter posits that it is precisely the uncertainty engendered by the ongoing ecological crisis that finally enables the blossoming of such intellectual multivalence around the idea of nature and the consequent development of qualitatively different approaches to relating – and, by extension, relating to – our complexly ecological reality and its contingent unfolding. In other words, instead of being treated as an obstacle to finding solace in the natural world, uncertainty may be somewhat counterintuitively approached as the oftentimes overlooked consolation in itself – the consolation of an open-ended more-than-human interrelationship – that the unsettled idea of nature promises in the face of the planet-encompassing degradation.

What the chapter aims to examine can, therefore, be summarised as twofold: what nature as a newly uncertain concept may mean and how its uncertainty may matter for developing a better – more ethical and more open-ended – approach to the other-than-human world. Specifically, what new concepts and imaginations of nature are emerging out of the present-day ecological uncertainty, and how engaging and, subsequently, reconciling with such uncertainty may enrich our understanding of the way the hopes and consolations traditionally associated with nature are transforming in response to the looming threat of environmental catastrophe. In search of some answers, I turn to the Anthropocene, this vibrant buzzword of a geological epoch, as a possible name for our troubled "here and now." Providing a rich theoretical framework for studying the suddenly tumultuous world, the term acts as an "intellectual shortcut" into its operations, an "expanded question mark" added to its rarely questioned presuppositions (Clark 3), and a material-semiotic source of its all-percolating uncertainty. All this makes it a fertile foundation from which one can start approaching both the roots and prospects of the recent ambiguity around the idea of nature. Most importantly, as the following section will argue, the Anthropocene's perspective allows for further inquiry into the inherent relationality of the idea of nature, with particular emphasis on the promising asymmetries and concomitant ethical intricacies of human-nonhuman entanglements.

2   The Self-Conscious Anthropocene: the End of Nature as We Know It

Since its conception in 2000, the Anthropocene has been summarily understood as the advent of the unprecedentedly drastic, because open-ended, rearrangement of planetary relations and our imageries thereof, highlighting the challenges of humanity's rapidly increasing impact on the other-than-human world as a perilous "telluric force" that unwittingly shapes and reshapes the Earth (Crutzen). The Anthropocene thus becomes not only a staggering realisation of how inherently ambiguous and conditional our conventional worldviews are but also an urgent call for developing radically new ways of comprehending the world and our increasingly confusing place in it.

In this regard, ecocritic Lynn Keller offers an insightful spin on the epistemologically disorienting recognition of the Anthropocene's manifold complexities by approaching them as the "self-consciousness" of the epoch: "a powerful cultural phenomenon tied to reflexive, critical, and often anxious awareness of the scale and severity of human effects on the planet" (2). For Keller, reconceptualising the awareness of the human-induced ecological degradation as the self-awareness of the Anthropocene challenges us to pay attention to how our seemingly objective epistemological practices are themselves always already implicated in the ontological rearrangements of the world, divulging such practices as inherently subjective, fragile, and, because of that, prone to change. Therefore, although the inherited Eurocentric onto-epistemology and its violently colonialist interpretative assumptions have always been part and parcel of the exploitation and devastation of the other-than-human world (Simpson), it should not necessarily be the only choice available. It is uncertainty engendered by the tumultuous transformations of the Earth that makes other onto-epistemologies possible, allowing for what Keller calls the "recomposing of contemporary consciousness [and] sensibilities" that troubles our thinking about the world beyond the human and frees it from the confinements of the convention (37). Therefore, the Anthropocene's onto-epistemological uncertainty opens up our unstable present as a field

of ethical possibility of embracing the epoch's liberating disorder and relating –
and thus always already relating to – the newly vulnerable and negotiable world
otherwise.

The present-day reckoning with the Anthropocene's complex uncertainties
can thus be summarised as, to quote literary scholar Pieter Vermeulen, "discourses and practices through which human anxieties and aspirations are artic-
ulated" (8). For Vermeulen, these "anxieties and aspirations" pertain primarily
to "the ongoing effort to find new names, images, and stories to make sense
of the bewildering changes to our sense of the relation between human and
nonhuman lives" (28). Not surprisingly, the ever-contentious idea of nature –
arguably, in its classical interpretation, the ideological pillar of Western onto-
epistemological anthropocentrism and its myopic certainty about the intrinsically hierarchical composition of the world (cf. Plumwood, *Environmental
Culture: The Ecological Crisis of Reason*) – remains at the very heart of the search
for new ways of imagining and navigating the convolutions of the Anthropocene's interconnectedness. More precisely, it underwrites the epoch's central
ethical conundrum: how to reconcile the recognition of human-nonhuman
inextricable entanglements with the need to preserve some kind of distinction
between the two, so that we can take responsibility for the disproportionate
severity of anthropogenic impact on the planet. While it is true that the revelations of the Anthropocene implore us to face, in Bill McKibben's notorious
formulation, "the end of nature" as "a separate and wild province" (48), the
perspective of the epoch's self-consciousness allows us not to fall prey to the
apocalyptic undertones and eschatological temptations of such recognition.
Instead, it allows us to approach the end of nature not as an ontological given
but as an epistemological one: the end of nature as we know it.

Since the Anthropocene's complicatedly more-than-human reality can be
accommodated by neither the conventional notion of nature nor the rejection
of this still-vital category altogether, we need to, following Donna Haraway's
urging, "find another relationship to nature besides reification, possession,
appropriation, or nostalgia" ("Otherworldly Conversations" 158). Thankfully,
out of the self-conscious Anthropocene's radical opening, there emerge new
definitions of what nature may mean and how it may matter in our ecologically turbulent times. Not only does the epoch's agential awareness "tak[e]
'nature' to be a far more inclusive and culturally imbricated category than has
been the case in our [i.e. American and Americanised] tradition of nature writing" (Keller 4), but it also, by blurring the boundaries of the hitherto exclusionary concept, may enable the blossoming of qualitatively new imaginations of
and attitudes towards the radically more-than-human world and its multifari-
ous contingencies. Geographer Jamie Lorimer even contends that "there is no

single Nature or mode of Natural knowledge to which environmentalists can make recourse" in the fruitfully "multinatural" Anthropocene (2), but instead "multiple natures are possible" (7). By embracing the epoch's uncertainties, one can thus finally, albeit belatedly, do justice to the sheer complexity of contemporary human-nonhuman interrelations with all their potent asymmetries and non-innocent symbioses.

Therefore, while the homogenised reification of Nature as pristine wilderness may indeed have ended, manifold uncertain natures of the Anthropocene sprout out of its ruins, both material and semiotic. The question remains, however, of how to relate to them in a manner that is simultaneously more attentive to their incongruous differences and more careful towards their delicate intertwinements. As the classical idea of human mastery over the nonhuman is being dismantled along with the reified concept of Nature, what better models of connecting with the wider world may arise in its stead? In my search for such models, I proceed from anthropologist Anna Tsing's intriguing declaration: "Ruins are now our gardens" ("Blasted Landscapes" 87). To explore what new, more accountable ways of knowing, imagining, and, hopefully, being-with the newly vulnerable more-than-human world may arise from the ongoing rearrangement of the human–nature interrelatedness, the chapter now turns to the imaginatively bountiful figure of the garden as a promising redefinition of the Anthropocene's onto-epistemological ruins of Nature as well as to the act of gardening as an ethical framework for both relating and relating to its uncertain, damaged, yet still promising natures.

3   Rambunctious Gardening: a Relational Ethics of Attentiveness
       and Care

The garden, either as a metonymic place or a metaphorical figure, may appear to be a fairly odd choice to fill the imaginative lacuna left in our imaginations of the other-than-human world by the advent of the self-conscious Anthropocene and its dismantling of the classical idea of Nature. Yet the figure of the garden may also be viewed as the site of ecological uncertainty *par excellence*: continuously fluctuating between nature and culture, wildness and tameness, design and randomness. All this makes it peculiarly suited for accommodating manifold perturbations of the Anthropocene's landscape, itself simultaneously material and intellectual. Moreover, as a proverbially indeterminate enterprise, the act of gardening may provide a useful template for learning not only how to coexist but also how to co-create with the dynamic uncertainty of the Anthropocene's ever-changing natures. Therefore, as this section intends to demonstrate, the complicated figure of the garden offers a unique way of attending to the otherwise elusive rearrangements of the present-day more-than-human world and a relational ethic of tending them towards a possibly better future.

While the trope of the garden as "second nature" quite obviously emerges as an alternative to the manifold drawbacks of the conventionally reified idea of Nature (Pollan), it still risks falling into the trap of the simplified, one-dimensional pastoral and its "outmoded [...] models of harmony and balance" (Garrard 65),[1] merely reinforcing the anachronistic vision of nature as some-thing static and unchangeable. Therefore, should the figure of the garden be useful for the challenges of the Anthropocene, it must be further problema-tised. Environmental journalist and thinker Emma Marris, for instance, argues that we have "misplaced nature" (1): striving for the unattainable standards of purity and authenticity exemplified by the reified vision of Nature, we have ignored the ongoing transmutations of the Anthropocene's messy and uncer-tain natures. Instead of a clear-cut answer to the trite question of what ought to count as nature and, therefore, ought to be worthy of our interest and pro-tection, Marris proposes to treat the whole epoch, with all its difficulties and contradictions, as a "global, half-wild rambunctious garden" in itself (2). Such a proposition opens up both nature and the definition thereof towards the unruly hybridity and dynamism of the world around us – from the sublime artificiality of national parks to the self-seeded ecosystems of our everyday-ness, where humans and nonhumans mix and mingle in the cracks of urban decay. Marris argues that this simultaneous redefinition of nature and gardens is both proactive and hopeful as "it creates more and more nature as it goes" (3), allowing us not only to dispel the dangerous fantasies of both pristine wilderness and paradisiacal pastoral, but also to fight against the debilitating pessimism of presumed "scarcity of nature" in our troubled times (McAfee). Approaching the Anthropocene's uncertain natures as rambunctious gardens broadens our perception of reality beyond the human, potentially leading to a better – more accountable and more fulfilling – relationship with the other-than-human world.

Most importantly, Marris's emphasis on the rambunctiousness of the Anthropocene's gardens distorts not only the illusion of pastoral orderliness

---

1   It must be noted, however, that Greg Garrard significantly expands and revises his views on the pastoral in the latest, third edition of *Ecocriticism* (2023), allowing for greater nuance and positive complexity in the term while remaining critical of its sentimental associations (40–64).

but also the figure of the human gardener who is supposed to maintain such an order. Whereas it may be tempting to embrace the unavoidability of human impact on the environment and rebrand our species as contemporary *homo hortensis*, such a "garden scenario" in which human mastery over nature is "total" yet supposedly "beneficent" would only lead to, in environmental historian Roderick Nash's felicitous formulation, "a planet-wide extension of Europe" and all the violent, colonial homogenisation associated with it (380–381). At their most questionable, such (mis)readings of the trope of the garden result in overly optimistic visions of the so-called "good Anthropocene" (Asafu-Adjaye et al.). There, in a subversion of Leo Marx's landmark study *The Machine and the Garden*, the hitherto dreaded machine is envisioned to become one with the Edenic garden, uncritically combining the worst of both worlds – technological utopianism and messianic pastoralism (Schleusener). In their more moderate renditions, such readings feed into the recent revival of the ethical concept of environmental stewardship (Bennett et al.) – at once heavily problematic, primarily because of its undertones of benevolent anthropocentrism, and much needed as it provides a gateway for the pressing debate around human responsibility and nature protection in the Anthropocene. By focusing on the entangled contingency of the epoch, Marris troubles both the conventional idea of gardening as well as the ethics of stewardship that come with it. In the Anthropocene, both humans and nonhumans share agential power: it is never quite clear who tends whom.

Yet, as argued in the previous section, by no means does the Anthropocene's onto-epistemological uncertainty translate into an ethical one. Therefore, even though the recourse to treating the epoch's rambunctiousness as merely another metaphor for the idea of "flat ontology" (Latour) – which substitutes the notion of nature with that of ecology understood as the radically horizontal interconnectedness of everything and everyone (Morton, *Ecology without Nature*) – is quite appealing as a possible absolution of the problem of anthropocentrism, Marris's focus on gardening does not shy away from the ethical conundrum of the Anthropocene's uncertain interrelatedness but prompts us to thoroughly engage with it instead. In this respect, Polish ecopoet and ecocritic Julia Fiedorczuk offers a scintillating figuration for understanding our place in the newly turbulent world and our role in its intricate disequilibrium – the idea of humanity being the "cyborg in the garden." Mixing Marx's aforementioned magnum opus with Haraway's iconoclastic "Cyborg Manifesto," Fiedorczuk's figuration elucidates how co-existing in the Anthropocene is always already about, in the words of Haraway, "pleasure in the confusion of boundaries" as well as "responsibility in their construction" ("Cyborg Manifesto" 7). And vice versa: how it is also about taking responsibility for

the uncontrollable, binary-defying uncertainty of the epoch while still finding pleasure and, by extension, consolation in tending its oft-unmanageable,
irreconcilable differences. Not only does such reformulation bring to the fore
the ticklish issue of accountability, thus resisting the temptations of both the
"good Anthropocene" and fairly naïve celebrations of planetary-encompassing
entanglements, it also highlights the multivalent act of gardening as an act of
the (de)construction of boundaries, at once material and semiotic, ontological
and epistemological, dangerous yet indispensable for differentiating between
and thus preserving the epoch's multitudinous worlds.

The process of gardening – as rambunctious as that of becoming the
more-than-human cyborg – may help us not only not get lost in the Anthropocene's sinuous interconnectedness but also cultivate a more responsible
approach towards its irreducible indeterminacy through two twinning facul-
ties typically associated with the heuristics of tending a garden: attentiveness
and care. Although attention and care have been conventionally linked to
the idea of human stewardship of the earth, geographer Anna Krzywoszynska asserts that the two should be considered as always already fostering the
much broader "more-than-human ethics" of co-inhabiting the Anthropocene.
Drawing on a plethora of theories of human-nonhuman interconnectedness,[2]
Krzywoszynska redefines fairly generalised "attention" as more context-
dependent "attentiveness" – "always an attentiveness towards" (665) – that
serves as a basis for the delicate relational ethics of care: "Practices of care
demand attentiveness to the entities cared for so that their needs can be known
and responded to" (672). Such relational ethics of attentiveness and care are
steeped in the Anthropocene's uncertainty: while "[a]ttentiveness introduces
uncertainty" to our assumptions about the "make-up" of the world (669), care
allows us to appreciate manifold possibilities opened by such onto-epistemological rearrangement – the possibilities of imagining and acting upon the
suddenly more-than-human world differently. Furthermore, by virtue of being
embedded, emplaced, and embodied in the actual process of gardening and
working the soil, the practices of attentiveness and care elucidate our complicated mundanity as the plane where, at the intersection of cripplingly nostalgic pasts and overly speculative futures, a meaningful change in the currently
ominous state of affairs may arise through the daily acts of "getting by, living

---

2   Among many others, the two figurations of attentiveness and care are of special importance
    for this chapter: Tsing's "arts of noticing" vulnerable more-than-human affinities haphazardly
    emerging out of the Anthropocene's ruins (*The Mushroom* 17) and Haraway's "response-ability"
    understood as an ability to adequately respond to the world's looming uncertainties and thus
    "cultivat[e] collective knowing and doing" in a more ethical manner (*Staying* 34).

alongside the world, living through it" in the pronouncedly "Anthropocenic present" (LeMenager 221).

On the whole, if we are to learn how to responsibly relate to our contingent world here and now, we need a relational ethic of rambunctious gardening. Attending to the elusive ways in which we are creating, and are being co-created by, the Anthropocene's disturbance of human-nonhuman interrelations, such ethics may teach us not only how to carefully engage with the epoch's prickly differences and fragile interlacements but also how to tend them towards a better future of affirmative reciprocity and beneficial coexistence. In the unruly messiness of the suddenly more-than-human world, there is a seedling of hope to be found and cultivated.

Coincidentally, the question of how to better relate to the Anthropocene's uncertain natures implies the question of how to better relate them, thus creating an imaginative task for literature to undertake. Literature itself can be considered as a garden of sorts: a medium where we can train our attentiveness to the emergence of new interrelationships and carefully respond to their ongoing metamorphoses. "That is the magic of both gardens and stories," writes literary scholar Robert Pogue Harrison, as "they transfigure the real even as they leave it apparently untouched" (95). In the remaining part, the chapter studies four texts (David Searcy's novel *Ordinary Horror*, Camille T. Dungy's memoir *Soil*, Robin Wall Kimmerer's treatise *Braiding Sweetgrass*, and Mei-mei Berssenbrugge's poetry collection *Hello, the Roses*) centred around gardens and gardening in order to examine how these selected texts of 21st-century U.S. literature reimagine the human–nature relationship in the Anthropocene. By weaving together these texts' highly variegated approaches to the rambunctious figure of the garden and its concomitant ethics, I hope to showcase what different, genre-specific literary strategies of attentiveness and care they employ in response to the epoch's looming uncertainties as well as how these strategies translate into a reconciliation with such uncertainties through a more appreciative understanding of the contingency and plurality of nature and the imaginations thereof.

4   "bloom how you must i say": David Searcy's *Ordinary Horror*
        and Camille T. Dungy's *Soil*

In the spirit of the Anthropocene's oftentimes surprising symbioses, this section commences its inquiry into the affordances of reimagining the epoch's uncertain natures through the figure of the garden by discussing Searcy's 2001 weird fiction *Ordinary Horror* and Dungy's 2023 memoir *Soil: The Story of a Black Mother's Garden.* Despite rather drastic differences in their techniques and perspectives, both texts trouble the homogenised vision of quotidian reality ignorant of the Anthropocene's ruinous progression by subverting the idea that the natural world, with all its disruptive wildness, is always out there and not in one's very own backyard. In paying attention to the sheer mundanity of the oft-overlooked rambunctiousness of human-nonhuman interrelationships, the two texts see a possibility of reimagining the present-day isolating status quo towards a more-than-human reality that is much more acceptable of contingencies and differences.

Described in a cover blurb as "a Stephen King novel written by Joseph Conrad," Searcy's *Ordinary Horror* ([2001] 2002) inventively mixes seemingly disparate genres and tropes – from the middle-brow psychological novel to the plant-horror trash story – to delve into what happens once the seemingly harmonious flow of day-to-day life is interrupted by the growing awareness of the wildly sprawling human–nature entanglements. There, septuagenarian widower Frank Delabano lives a reclusive life somewhere on the brink of the Great Plains in a dilapidated suburban tract house, whose enclosed "ordinariness" he cherishes for "seem[ing] more or less permanent" (1), and tends his loneliness by tending a rose garden. Delabano "loves [roses] springing from such ordinariness" (1); "their spikes pinning things in place" (5), "filling the world with predictable phenomena" (16). Everything changes, however, when his garden becomes invaded by gophers and Delabano orders a pest-repellent plant, "exotic, South American, never-before-available," guaranteed to be "antithetical to garden varmints but harmless to pets and everything else" (3–4). Once the allegedly "nonflowering" plant prepares to bloom (3), the seemingly permanent order of the everyday, along with the carefully manicured garden, begins to collapse into the chaos of uncertainty.

An avid gardener above all, Delabano "hesitates to interfere" into the plant's blossoming "for fear of disrupting something," feeling that "he should keep his distance, let things happen" (27). When the plant finally blooms "huge, blue roses" (28), unruly strangeness erupts out of the ordinary: be it the cicadas that, out of the whole neighbourhood, appear to invade only his backyard, or the indescribable yet eerily fascinating smell that suddenly floods his house. Delabano, grappling for meaning, immediately links these extraordinary occurrences to the plant's jungle origin, "primitive and strange" (17), and its allegedly corruptive agency that leaves his small picket-fenced world of suburbia "permeable and hopeless" (74). "[W]anting reassurance, reconciliation with the ordinary facts" (29), he looks for some kind of rational explanation but finds only a sole mention of the cryptic plant in a suspiciously unchecked library book entitled *Amazonian Biotoxins* (89) – an occult colonial-era travelogue and an explicit pastiche of *Heart of Darkness* – and, for a while, becomes obsessed with the idea of explaining the unfolding weirdness through the time-tested frontier horror story of the wilderness's encroachment onto the garden's peaceful predictability. Yet, as Delabano carefully attends to flickering lights, rattling noises, and gnarled shadows for most of the poetically winding novel, this is the suburbia that slowly reveals itself to be inherently sinister and despairing: from the TV-bound silent spinsters next door to the new neighbours' neglected child that keeps sneaking into his house. When contrasted with the ordinary horrors of Delabano's mundanity, the strangeness of the plant becomes less dreadful and more liberating, spelling with its apparent otherworldliness the possibility of an alternative to the suffocatingly claustrophobic suburbia, with all its haunting remnants of the frontier mentality, its intentional severance of any meaningful connection with the other-than-human world, and the resulting numbing loneliness of its ghost-like inhabitants.

This realisation leads to the fifty-pages-long culmination of the novel where, during an especially esoteric "barbecue Friday" (183), Delabano's neighbours, becoming more and more inhuman in their appearances and behaviour, confront him about his reluctance to get rid of the invasive exotic until the whole suburb disintegrates in "green flames," leaving behind the openness of the prairie, "as if almost nothing were there to begin with [...], the less substantial the more combustible, a sort of miracle for everything to have lasted so long" (230). Intentionally leaving the reader confused, the novel's climax is deeply ambiguous but not at all relativistic. Its strategic uncertainty cannot be reduced to some realistic explanation (Delabano's alleged senility or, indeed, the hallucinogenic influence of biotoxins), nor can it be entirely explained away as a mere metaphor or a supernatural fantasy. Instead, the climax's addled revelation hovers somewhere in between. On the one hand, it makes us, along with Delabano, finally notice the prairie's materiality irreducible to the colonialist fantasies of malicious horrors. On the other, it dismantles the assumed permanence of our business-as-usual everydayness as a mere shared game of pretence, leaving in its place "a sticky, grainy feeling" of thickening onto-epistemological uncertainty (111). Searcy's novel thus illustrates how it is the exclusionary binaries through which we tend to uncritically approach the world – like the pastoral/wilderness dichotomy of the good and civilised garden versus the wild and horror-inducing jungle – that are, in fact, preternatural, especially considering their helplessness when confronted with the sheer intricacy of today's ecological reality.

Written at the very nascence of the self-conscious Anthropocene, *Ordinary Horror* does not offer any affirmative model of more-than-human coexistence to supersede the conventionally binary approach to nature, yet it quite presciently demonstrates how attentiveness to the problematic cracks in our habitual worldviews may lead to a weirdly consoling realisation: not only are other imaginations of human-nonhuman interrelationship possible, they are always already lurking behind the façade of our presumed certainties. Most interestingly, Searcy's novel appears to present uncertainty as a crucial onto-epistemological condition out of which more ethical ways of relating to the other-than-human world may emerge. Looking for some particular examples of such qualitatively different relation to nature enabled by uncertainty, this section now turns to Camille T. Dungy's much more recent and thus much more ecologically aware memoir *Soil* (2023) that continues Searcy's interrogation of the not-so-ordinary experiences of nature in the Anthropocene but takes it to a direction that is pronouncedly conciliatory, both towards the manifold uncertainties that surround nature and the mundanity of the American suburbia.

Spanning seven years, one global pandemic, and the worldwide surge of the Black Lives Matter protests, Dungy's experiential memoir *Soil* recounts a story of discovering a new relation to our radically interconnected world through tending a garden in a predominantly white suburban neighbourhood of Fort Collins, Colorado. Against all odds – the local homeowners' association's self-righteous crusade against any "unwieldy or weedy vegetation" (43), the all-too-high "physical and mental demands of a robust garden" (32), and the "seeming improbability of a Black mother writing about nature" (62) – Dungy engages in turning her backyard into a part of the Prairie Project, a wildlife restoration endeavour that encourages Midwestern gardeners to reintroduce native species into their lawn-and-concrete mundanity. In stark contrast to the ambiguous indirectness of Searcy's novel, Dungy's memoir openly depicts the seemingly oxymoronic practice of wildlife gardening as an inherently political act of "stay[ing] open to surprise," of "stay[ing] open to lives that look and act in radically different ways than [one is] used to or comfortable with" (47). In the paradox of carefully tending the rambunctiously heterogeneous and thus generatively contingent garden, Dungy sees a way of "resistance" to the "suburban American monoculture" of certainty, indifference, and estrangement (44). Therefore, her attentiveness to the inherent hybridity of nature as well as her reluctance to fully assume the role of the all-controlling gardener become the foundation for cultivating a more-than-human polyculture of blurred boundaries and affirmative coexistence.

Although Dungy sees the suburban monoculture to be primarily exemplified by the homogenised sterility of the artificially harmonious pastoral that shuns any difference, either ecological or cultural, in order to maintain its fantasy of stability, she, unlike Searcy's Delabano, does not imbue the ordinary with sinister undertones. Instead, she recognises it as the only plane where one can actually experience and appropriately respond to "[t]he instability that is the only stable truth" about the Anthropocene's human-nonhuman interconnections (137). Simultaneously intensified and distorted by the COVID-19 isolation, the laborious daily routine of taking care of her home, family, and garden allows Dungy to appreciate not only how rich the other-than-human world of her backyard is but also how deeply her well-being is dependent on it. Crucially, such recognition also makes her notice that the devastating monoculture she so vehemently opposes is not limited to the HOA-mandated idyll but is also propagated by the idea of similarly monolithic Nature, all too often regrettably depicted as, in Dungy's formulation, "someplace separated from the action of everyday lives" (138). Namely, by a certain "pattern in nature writing that confounds and annoys" her (66): be it the allegedly self-sufficient machismo in the manner of Edward Abbey, whose self-imposed survivalism is devoid of any mentions of his wife and family who cared for him during his "solitary" stay in the Colorado Plateau wilderness (Abbey [1968] 2020), or the detached epiphanies in the vein of Annie Dillard, whose yearning for the truly transcendental communion with nature intentionally omits any sublunary matters and concerns (Dillard [1974] 2011). The problems with the aforementioned pattern can be encapsulated by a simple yet valid question Dungy poses: "Why doesn't anyone in fundamental environmental literature seem to have to do the dishes?" (85).

Not only does such disinterest in more down-to-earth concerns deepen the separation between the experiential plane of ordinary life and conventional imaginations of nature, but it also leads, as Dungy later argues, to the dangerous "separation between the environment and social justice" (212). For her, the ability to completely uncouple the daily troubles of one's life from one's contact with nature is a privilege inaccessible to the majority of us who live in the ecologically complicated world of pervasive water and air pollution and zoonotic pandemics. If left unacknowledged, such privilege, which has for decades dominated U.S. nature writing, risks glossing over the sheer plenitude of socio-cultural contexts along with the respectively differentiating variety of our interrelationships with the other-than-human world, sometimes positive, sometimes negative. It is precisely Dungy's unwillingness to disentangle quotidian intertwinements of her lived experiences – as a Black mother, an unorthodox nature writer and poet, and simply someone who tries to navigate the uncertainties of the 21st-century world with mindfulness and care – that allows her to avoid the trap of extolling the reified idea of Nature, a mere reversal of Delabano's colonialist chimaera of the horrific jungle, as a viable alternative to the equally exclusionary confinements of the suburban pastoral.

Instead, Dungy argues that these two U.S. monocultures – wilderness and pastoral – should be perceived as twinning ideologies of "culturally constructed weeding" that are defined by their desire to eliminate any uncertainty about where the human world ends and the natural one begins (44). Weeds themselves become the focus of Dungy's gardening endeavour once the local HOA deems her rambunctious wildlife backyard to be an instance of gross misconduct and altogether too "aesthetically unsavoury" (43). This allows Dungy to notice the inherent arbitrariness of our seemingly commonsensical assumptions about which plants are worth tending and which are not. Furthermore, Dungy links the cultural construction of weeds to a much broader "violence of our imagination" rooted in the original Western monoculture (174) – the violence of the plantation – which Jan Beneš discusses in greater detail later in this volume – and its demands for utmost effectiveness, purity, and subservience. Contemplating the uneasy link between gardens and plantations in the light of the George Floyd protests, Dungy's husband even remarks: "We're like dandelions. When they see us, they try to kill us" (119). This prompts Dungy to further sympathise with her weedy plot and its alleged uselessness, messiness, and unruliness. Feeling affinity with her garden's subversion of the colonial axiology of the plantation, Dungy learns not only how to accept but also how to cherish its imperfections for their ability to demonstrate how it is the contingent hybridity of human-nonhuman interrelations that deserves our utmost attentiveness and care.

By attending to the defiant rambunctiousness of her garden that she so carefully documents (in trial-and-error anecdotes, photographs of garden plants and animals, both indigenous and not, and occasional poetry), Dungy comes to a conclusion that "life grows more complex when we muddle hierarchies of power" (163), that "[s]ometimes unmastered growth reveals our dearest needs" (211) – including the need for a reconciliation with the world's irreducible uncertainty that has the potential to subvert the violently self-perpetuating cycle of Western monocultural (mis)interpretations of nature either as the pristine wilderness (treated as the jungle to be bravely conquered) or the idyllic pastoral (treated as the plantation to be constantly managed). That is, the need for a fulfilling human-nonhuman relationship that would allow us to imagine what a new, better, "greater-than-human" world might be (179). This revelation is best summarised by a vivid line from African American poet Lucille Clifton that Dungy adopts as her memoir's epigraph: "bloom how you must i say."[3] In this imperative to "relinquish control" (173), to "see [...] more abundantly" (151),

---

3   The line comes from Clifton's poem "mulberry fields" which reckons with the hauntings of the irreparable trauma of slavery plantations. The poem can be found in the trailblazing

there is a more-than-human ethic for the Anthropocene to be glimpsed. In such an ethic, attentiveness to the cracks in the American monocultures grows into active care for what uncertain natures are already wildly springing out of their ruins – care for hope, not horror. If Searcy's novel hints at the possibility of a strange solace of a meaningful more-than-human relation hidden behind the anthropo- and Eurocentric monocultural imaginations of nature, Dungy's memoir, along with her thriving wildlife garden, showcases how this solace may be found in the appreciation and cultivation of the very uncertainty that generatively destabilises such imaginations and their assumed hierarchies.

5   "The land a kind of distributed feeling": Robin Wall Kimmerer's
       Braiding Sweetgrass and Mei-mei Berssenbrugge's *Hello, the Roses*

If the previous section was preoccupied with elucidating how the open-ended practice of gardening, when situated in the context of the U.S. suburbia, can trouble the homogeneously reified idea of Nature along with the colonial worldview behind it, this section is more interested in what qualitatively different ways of knowing and of being-with may instead be developed through the ongoing redefinition of the Anthropocene's human-nonhuman interrelationship as the planet-wide rambunctious garden. By weaving together another non-obvious literary dyad – Kimmerer's 2013 non-fiction *Braiding Sweetgrass: Indigenous Wisdom, Scientific Knowledge, and the Teachings of Plants* and Berssenbrugge's 2013 poetry collection *Hello, the Roses* – this section intends to examine what role the politics and poetics of gardening may play in creating a new relational ethic for the Anthropocene. Specifically, an ethic of tending our uncertainly unfolding present towards a truly communal more-than-human future – the future of affirmative cohabitation and reciprocal, albeit asymmetrical, cultivation.

Written as a "healing" exercise in "intertwining [...] science, spirit, and story" in response to the human-induced deterioration of the other-than-human world in the Anthropocene (Kimmerer x), Kimmerer's *Braiding Sweetgrass* is indeed an intricate assemblage of genres dedicated to reconciliation with the epoch's confusing entanglements and potent incongruities: an Anishinaabekwe's autoethnographic memoir of integrating Indigenous myths into one's Americanised mundanity, a botanist-cum-poet's field notes on shortcomings of the Eurocentric scientific method, a political manifesto for a vivaciously

---

anthology *Black Nature: Four Centuries of African American Nature Poetry* (2009) compiled and edited by Dungy (260).

multispecies community, and a deeply personal essay in defence of the imaginative power of language to bind such a community together. What unites these
seemingly disjointed perspectives and sets their unlikely interweavements into
motion is Kimmerer's book-long search for a new cosmology, a new "source
of identity and orientation to the world" (7), that would replace the Western
onto-epistemology whose claim to the universally objective worldview, based
on the belief in "science" that "separates the observer and the observed" (45),
becomes especially questionable when faced with the Anthropocene's rambunctious, binary-defying uncertainties. Instead, Kimmerer argues for a vision
of a "universe that is a communion of subjects, not a collection of objects"
(56) – a cosmology that, on account of being a truly open-ended, uncertainty-
and difference-embracing orientation, would bring about what she calls a
"re-story-ation" of the human–nature interrelationship (9).

Drawing on the Native American creation myth of Skywoman who originated Turtle Island as "a garden for the well-being of all" (6–7), Kimmerer situates the gestation of such a new cosmology in the figure of the garden which
she sees as the "nursery for nurturing connection" (126), a practice that is at
once material and spiritual. In the generative tension between the dynamic
materiality of her private garden as a world of its own – on which Aleksandra
Kamińska elaborates earlier in this volume – and the spirituality of approaching the whole planet as an interconnected multitude of "gardens, known by
some as 'global ecosystems'" (6), Kimmerer finds a way to reconcile our world's
sheer ecological abundance with the Anthropocene's ethical imperative to
reclaim responsibility for such world's uncertain unfolding. A template for
such a reconciliation comes from the sacred Indigenous ceremony of planting, tending, picking, braiding, and burning – in other words, gardening – the
eponymous sweetgrass. During her botanical study of the possible reasons for
the plant disappearing from its traditional habitats, Kimmerer discovers that
the sweetgrass "has apparently become dependent on humans to create the
'disturbance' that stimulates its compensatory growth" (164): defying the conventional predictions of her colleagues, the endangered plots were not those
traditionally harvested by the local Indigenous communities but those left
untouched in the hope of preserving the species. This leads her to a conclusion
that, since a fruitfully troubling human interference is oftentimes needed, the
whole "work of being a human" actually lies in "finding [such appropriate] balance" of symbiotic trouble (146), "so that attention" towards the world's entangled rambunctiousness "becomes intention" to ethically reciprocate its gift of
interconnection (249).

As it is Indigenous land-care practices that may teach us how to find the
much-needed balance, they ought to be protected alongside the endangered

wildlife they helped to co-create. For "wild ideas are in jeopardy" (201), too. For Kimmerer, the Western onto-epistemology's homogenising imperilment of rambunctious Native American cosmologies is primarily twofold. On the one hand, it is revealed in Western science and its intentional separation of knowledge-producing practices from ethical concerns: "Science can give us knowing, but caring comes from someplace else" (346). On the other hand, it is represented by the present-day reliance on large-scale agriculture – as exemplified by the U.S. "monoculture of corn in straight rows of indentured servitude" (175) – that estranges us from both the possibility of spiritual connection with the other-than-human world and our ethical responsibility towards it as co-creators. Therefore, not only should "the intellectual monoculture of science [...] be replaced with a polyculture of complementary knowledges" (139), the modern society based on industrial agriculture should also be superseded by a more ethical community that would give us both a reason and a possibility to care. Such a planet-wide "democracy of species, not a tyranny of one" (58), would be founded upon the indigenous gift economy, its currency of reciprocity, and the ethics of troublesome gardening – upon a recognition that "[a]ll of our flourishing is mutual," albeit unequal (166). Most importantly, this would allow us to finally approach the surrounding world as the ultimate responsibility – our very human obligation to balance out its differences and contingencies without explaining them away.

Whereas this vision of the world to come is obviously speculative, it is not at all just wishful or fatuous. As Kimmerer argues: "It's not enough to grieve. It's not enough to just stop doing bad things" (328). Yet, in the uncertain times of the Anthropocene, to start doing good things, one needs a vision, a cosmological orientation, a wild openness of possibility: "[O]ur work" is "to discover what we can give" (239). It is in the uniquely human faculty of imagination that Kimmerer sees such an opening. For her, "we are storymakers, not just storytellers" (341), and language is our main gift and our greatest responsibility to the more-than-human world. We can thus reciprocate the world's boon of interrelationships by learning how our "uncertain path to the future can be illuminated" by wildly hopeful fabulations, veritably cosmological in their scope and ambition, that we should create to disturb the intellectual status quo (383). That is, by learning what Kimmerer, recalling her own experience of reclaiming the Potawatomi language, calls the "grammar of animacy" (48) – an ethic of relating and thus always already relating to the ever-differentiating more-than-human world with attentiveness and care – we may be allowed not only not to "pas[s] the fork in the road" towards a healed communal future but imagine a better vision of the future altogether (371). With her multi-generic treatise, Kimmerer appears to offer such a literary vision of a story-made world where the desired democracy of species is represented through the veritable democracy of genres and imaginations of nature that thrive and tangle in their generative multivalence.

To further flesh out Kimmerer's material-spiritual project of cultivating a wholesome multispecies community as well as to somewhat complicate her fairly utopian understanding of language as a conduit for bringing this community to life, I would like to end my inquiry into uncertainties surrounding the human–nature interrelationship in the self-conscious Anthropocene by turning to Mei-mei Berssenbrugge's *Hello, the Roses* (2013), a meticulous poetic study of how awareness blurs with uncertainty and uncertainty blooms into connectivity.

Berssenbrugge is a Chinese American poet whose experimental poetics of "contingency and relationality" (Wang 244), of "radical attention, observation, description" (Leong 41), has been universally acclaimed for its masterful translation of embodied and embrained phenomenological experience into verbal form. In *Hello, the Roses*, Berssenbrugge for the first time directly applies her well-established poetics to the intricate reality of the Anthropocene, embedding and emplacing her intimate experiences thereof in the "polyvalent" figure of the garden (30), hovering uncertainly between the fleeting impressions of her garden in New Mexico, destroyed in real life by an abnormal drought, and the growing awareness of the planetary "upheaval of many beings participating over a long time that appears like her time in the plot" (9). As it happens, this is also one of the very few of Berssenbrugge's poetry collections that are not aligned horizontally to accommodate her signature landscape-oriented lines. Instead, although the lines remain the same, the slender physical copy of the collection is strikingly elongated, "like an open window through which the moment is perceived" (17), not landscape- but garden-focused, as if to ensure that the reader adjusts their optics to the everyday and the minuscule. And yet, "I call it a vase," Berssenbrugge writes, "upward space for materialising flowers or birdsong where openness is form" (43). It is in renegotiating this "resonance of disjunction" between focus and openness, awareness and uncertainty, mundanity and cosmicity that her book finds its imaginative drive, linguistic exuberance, and ethical imperative (45).

For Berssenbrugge, who suffers from pesticide-induced hypersensitivity, virtually any contact with the outer world is conceived as an experience of utmost vulnerability – of being "open to accident" (3), to "a spontaneity of perception" (67). To be truly aware of one's environment is to be continuously overwhelmed with the kaleidoscope of synesthetic impulses assaulting one's senses, "with impressions like fingerprints all over" (59). Liberated from the confinements of sensory discrimination, this ecological hyperawareness begets a vertiginous feeling of uncertainty about where the human ends and the nonhuman begins: "There's no stopping this effusion. // Looking at the plant releases my boundaries" (51). Instead of being lamented as estranging or paralysing, however, such boundary-defying vulnerability is approached as promising a possibility of communion: "It defines my capacity to perceive a garden as if of its same nature" (56). In *Hello, the Roses*, Berssenbrugge is concerned with tracing how a more-than-human community may emerge out of the shared feeling of radical openness, while also remaining attentive to the still-insuperable "resistance of a [nonhuman] phenomenon to reveal itself" to human senses (51). As the collection's whimsical title suggests with its neighbourly greeting, she, similarly to Kimmerer, sees language that transcends the limitations of the human as a plane on which such an emergence may occur: "I dream all plants and animals communicate" (23). Still, while Kimmerer expects language to provide an orientation, Berssenbrugge is much more interested in its disorienting powers that, "cohering attention and feeling" (50), may somewhat paradoxically elucidate how "environmental uncertainty" already "act[s] as imagination of the group" (33) – of the community understood as the fragile reciprocity of human-nonhuman (mis)communication. To put it differently, she pursues the ethics of Kimmerer's grammar of animacy by carefully attending to the inherent animacy of grammar.

A Language poet and a non-native speaker of English, Berssenbrugge has always been fascinated by the treacherous agency of language to shape and reshape the way we relate to the world by suddenly making us aware of and thus uncertain about the conventional ways we relate it. She writes: "Our perceived world under the blue has cosmological interiority, as the relation of the whole cosmos to description, and characters breathe within a voluminous story" (81). In *Hello, the Roses*, Berssenbrugge employs her highly idiosyncratic style to demonstrate how such "cosmological interiority" may be revealed through careful observations of routine human-nonhuman interactions: "Communication flows back and forth between the rose and myself, and I begin perceiving through the plant [...] a touch from the world carrying experience all at once" (57). While her multipart poems with their iterative one-sentence lines alert the reader that "[a]n experience is never one experience" (27), her snaking syntax and bumpy punctuation serve "to concentrate presence" of uncertainty in the poem (90), helping the reader to develop appreciation for the ineffability of nonhuman rambunctiousness that normally exists outside the grasp of our blinkered cognition but can nevertheless be sensed through the arresting ambiguity of language "as appearance, word, reference, entangling event" (79). Combined with her capriciously capacious vocabulary that playfully mixes academese with quantum panpsychism, these poetic strategies allow Berssenbrugge to compose a "voluminous story" about how the "power with which attunements arise, the power of poetry, animates objects in heaven, rainbows, daystars, wild rose" (68), "transform[ing] uncertain space" of a superficially ordinary garden "into cosmos" whose "sacred[ness] is saturated with being" of contingently sprawling more-than-human entanglements befitting the Anthropocene in their scope and scale (92).

In its contradictory poetics of contrived serendipity of associations (from an ordinary "wild rose" to Dantesque "daystars" and from the particular garden to the whole cosmos), *Hello, the Roses* offers a model of ethics for the Anthropocene's relational disarray based on the collectivity of human-nonhuman exchange – an inkling of how Kimmerer's idea of storymaking a better world is always already a collaborative practice where "[d]ifferent species communicate and energies of environment and inhabitants merge" (31). For Berssenbrugge, it is not enough, however, to merely describe the radical yet oft-ambivalent connectivity of our turbulently ecological world. Instead, it is the poet's responsibility to imbue such connections with meaning, to become a gardener of sorts that tends a shared "sense of place as [the] continual emergence" of the more-than-human community out of the ever-uncertain yet still-present *inter*connectedness of open-ended communication (83). Her poetry collection can thus be read as a tongue-in-cheek slant on gardening manuals that, by provoking overwhelming attentiveness to the wider world, teaches us persistent care for its incongruous intertwinements. For it is also our responsibility as readers to notice, along with Berssenbrugge, how, once the irreducible uncertainty of the Anthropocene is embraced, a "body bec[o]me[s] a kind of distributed thing and the land a kind of distributed feeling," and to respond to this complex feeling that "is like a keyword or letter representing [one's] experience and also the continuum from individual to a consensus" (84). The feeling can be understood as that of multispecies solidarity that may "rise over us, like a direction or corridor" to a better future (84), should we just learn how to recognise and cherish the sudden vulnerability of our inherited Eurocentric onto-epistemology and the promising openness of boundaries, at once material and semiotic, that comes with it.

Crucially, as Berssenbrugge's poetry collection appears to argue through its unabashed literariness, it is also the responsibility of literature as such – this applied imagination of ours – to teach us how the poetically intensified "unbounded ambiguity" of the Anthropocene can be translated into politically charged solidarity that "reverses and recovers" (93) our hitherto destructive relationship with other-than-human beings by re-versing our commonsensical imaginations of nature in the subverting rambunctiousness of language and thus re-covering their dangerous binaries with boundary-blurring uncertainties of contingent associations. Contrary to some popular anxieties surrounding its seeming impotence in the face of the anthropogenic ecological breakdown, literature is revealed to be a potent, because prickly, and hopeful, because uncertain, meaning- and connection-making practice. Literature's value in the Anthropocene thus seems to lie in its unique command of uncertainty that, by keeping the aforementioned "direction or corridor" of multispecies solidarity contingent and ever-forking, makes us much more attentive to and careful about how we "cultivate inter-being" with the natural world and articulate its nascent alternative cosmologies (61). If Kimmerer, with a refreshingly unapologetic optimism, posits imagination as a faculty through which a better more-than-human future may come to fruition, Berssenbrugge cautiously demonstrates how language as an inherently uncertain medium may enable this: through fostering a powerful feeling of onto-epistemological uncertainty that, with attentiveness and care to its inherent hopefulness, may translate into an ethic of imagining the truly cosmic coexistence and multispecies solidarity.

6   In Lieu of a Conclusion: Consolations of Uncertainty

Having started with the question of what the newly volatile idea of nature may mean and how its uncertainty may matter in the self-conscious Anthropocene, I end my inquiry by circling back to Haraway's irreverently enthusiastic conclusion, which the chapter adopted as its epigraph: "We have, finally, no clear or distinct ideas" ("Situated Knowledges" 596). Instead of quibbling over some authoritative definition of what should count as nature to preserve and search for solace in, which would, admittedly, only curtail its imaginative potential, the chapter tried to demonstrate how the notion's seemingly troublesome uncertainty may in itself be an oft-overlooked and thus imaginatively invigorating consolation the notion offers – the multivalent consolation of the openness of possibilities of relating and relating to the other-than-human world otherwise. Consequently, the chapter argued that it is our responsibility to keep, in Haraway's famous urging, "staying with the trouble" as it is in learning how to reconcile with the trouble of the Anthropocene's relational mess, where the human and the nonhuman worlds twist and turn, that we may find a much-needed solace in our tumultuous times.

Thanks to the generative differences and disruptive affinities between their capacious attitudes and imaginative practices, not only do the four texts collectively provide an appropriately complex representation of the Anthropocene's labyrinthine uncertainty, but they also present a variegated array of understandings of what consolations such uncertainty may afford. Searcy's novel utilises its horror-story strategies to elucidate how ecological uncertainty may trouble the seemingly commonsensical readings of nature as either wilderness or pastoral and thus reveal itself to be strangely liberating from their binary confinements, providing a solace from the homogenous and homogenising Eurocentric onto-epistemology. Dungy develops this insight further by focusing on how reconciling with the contingency that the Anthropocene brought to one's situated mundanity may translate into a more meaningful personal connection with nature: for her, the main consolation the epoch's uncertainty offers is that of the definition- and hierarchy-defying rambunctiousness of the other-than-human world. Extrapolating this revelation from the personal scale to the planetary one, Kimmerer passionately argues that the blurriness of the conventional onto-epistemological binaries caused by the advent of the self-conscious Anthropocene enables the emergence of qualitatively different cosmologies, thus finally highlighting the solace of the more-than-human community and reciprocity the appreciation of which has for generations existed among the Indigenous cultures. Finally, for Berssenbrugge, the primary consolation of the Anthropocene lies in the potent multivalence of its sprawling discourse that, through the sheer indeterminacy of its free associations, results in the heightened attentiveness to how the epoch's entanglements, simultaneously material and semiotic, are always already shaping and reshaping our reality.

Ultimately, as the chapter's analysis of the four texts hopefully managed to explicate, it is by carefully attending to the ongoing material-semiotic transformations of nature that we may learn how to meaningfully coexist and co-create with the Anthropocene. In so doing, we must learn how to see ourselves not as helpless witnesses to the unfolding spectacle of ecological catastrophe but, instead, as the epoch's very gardeners: at once, as Searcy's novel so inventively demonstrates, its inadvertent perpetrators responsible for its already unfolding perils, and its reluctant stewards responsive to its tentative promises. From this perspective, the Anthropocene reveals itself to be not unlike the now-proverbial Borgesian garden of forking paths that nevertheless remains first and foremost a garden whose uncertain natures, as Dungy would arguably agree, require our utmost attentiveness and care. Therefore, it is also our responsibility to develop a new ethic of living with and through the Anthropocene, so that we may not only reconcile with its persisting contingencies but also learn how to cultivate these contingencies towards a better – more inclusive, more accountable, and simply more hopeful – vision of our inescapably ecological future. Still, before coming to life, such a vision needs to be, as Kimmerer would have it, imagined and figured out of the Anthropocene's promising open-endedness. It is literature, as Berssenbrugge's poetry collection seems to imply, that may teach us how to navigate the cacophonous chaos of ever-differentiating genres, poetics, narratives, metaphors, perspectives, and feelings which is the Anthropocene without diminishing its potent multivalence and thus foreclosing a possibility of finding a solace of a mutually beneficial, albeit still asymmetrical, more-than-human coexistence and co-creation.

Bibliography

Abbey, Edward. *Desert Solitaire: A Season in the Wilderness*. 1968. William Collins, 2020.
Asafu-Adjaye, Linus Blomqvist, Stewart Brad, et al. "An Ecomodernist Manifesto." *www. ecomodernism.org*, April 2015, pp. 1–32. Available from: https://www.ecomodernism .org/s/An-Ecomodernist-Manifesto.pdf.
Bennett, Nathan J., Tara S. Whitty, Elena Finkbeiner, et al. "Environmental Stewardship: A Conceptual Overview and Analytical Framework." *Environmental Management*, vol. 61, 2018, pp. 597–614. https://doi.org/10.1007/s00267-017-0993-2.
Berssenbrugge, Mei-mei. *Hello, the Roses*. New Directions, 2013.
Clark, Timothy. *Ecocriticism on the Edge: The Anthropocene as a Threshold Concept*. Bloomsbury Academic, 2015.
Crutzen, Paul J. "Geology of Mankind." *Nature*, vol. 415, no. 23, 2002, n. p. https://doi .org/10.1038/415023a.
Dillard, Annie. *Pilgrim at Tinker Creek*. 1974. Canterbury Press, 2011.
Dungy, Camille T. *Black Nature: Four Centuries of African American Nature Poetry*. University of Georgia Press, 2009.
Dungy, Camille T. *Soil: The Story of a Black Mother's Garden*. Simon & Schuster, 2023.
Fiedorczuk, Julia. *Cyborg w Ogrodzie: Wprowadzenie do Ekokrytyki*. Wydawnictwo Naukowe Katedra, 2015.
Garrard, Greg. *Ecocriticism*. 2004. 2nd ed., Routledge, 2012.
Garrard, Greg. *Ecocriticism*. 2004. 3rd ed., Routledge, 2023.
Haraway, Donna J. "A Cyborg Manifesto: Science, Technology, and Socialist-Feminism in the Late Twentieth Century." 1985. *Manifestly Haraway*. University of Minnesota Press, 2016, pp. 3–90.
Haraway, Donna J. "Otherworldly Conversations, Terran Topics, Local Terms." 1992. *Material Feminisms*, edited by Stacy Alaimo and Susan Hekman, Indiana University Press, 2008, pp. 157–187.
Haraway, Donna J. "Situated Knowledges: The Science Question in Feminism and the Privilege of Partial Perspective." *Feminist Studies*, vol. 14, no. 3, Autumn 1988, pp. 575–599. Available from: https://www.jstor.org/stable/3178066.
Haraway, Donna J. *Staying with the Trouble: Making Kin in the Chthulucene*. Duke University Press, 2016.

Harrison, Robert Pogue. *Gardens: An Essay on the Human Condition*. Chicago University Press, 2008.

Keller, Lynn. *Recomposing Ecopoetics: North American Poetry of the Self-Conscious Anthropocene*. University of Virginia Press, 2018.

Kimmerer, Robin Wall. *Braiding Sweetgrass: Indigenous Wisdom, Scientific Knowledge and the Teaching of Plants*. 2013. Reprint ed., Penguin Books, 2020.

Krzywoszynska, Anna. "Caring for Soil Life in the Anthropocene: The Role of Attentiveness in More-Than-Human Ethics." *Transactions of the Institute of British Geographers*, vol. 44, no. 4, 2019, pp. 661–675. https://doi.org/10.1111/tran.12293.

Latour, Bruno. *Reassembling the Social: An Introduction to Actor-Network-Theory*. Oxford University Press, 2005.

LeMenager, Stephanie. "Climate Change and the Struggle for Genre." *Anthropocene Reading: Literary History in Geologic Times*, edited by Tobias Menely and Jesse Oak Taylor, The Pennsylvania State University Press, 2017, pp. 220–238.

Leong, Michael. "Traditions of Innovation in Asian American Poetry." *The Cambridge Companion to Twenty-First-Century American Poetry*, edited by Timothy Yu, Cambridge University Press, 2021.

Lorimer, Jamie. *Wildlife in the Anthropocene: Conservation After Nature*. University of Minnesota Press, 2015.

Marris, Emma. *Rambunctious Garden: Saving Nature in a Post-Wild World*. Bloomsbury Publishing, 2011.

McAfee, Kathleen. "The Politics of Nature in the Anthropocene." *RCC Perspectives*, no. 2, 2016, pp. 65–72. Available from: https://www.jstor.org/stable/26241360.

McKibben, Bill. *The End of Nature: Humanity, Climate Change, and the Natural World*. 1989. Revised and updated ed., Bloomsbury Press, 2003.

Morton, Timothy. *Ecology without Nature: Rethinking Environmental Aesthetics*. Harvard University Press, 2007.

Nash, Roderick Frazier. *Wilderness and the American Mind*. 1967. 4th ed., Yale University Press, 2001.

"nature." *Oxford English Dictionary*. Oxford University Press, revised 2003, no pagination. https://doi.org/10.1093/OED/1850732511.

Plumwood, Val. *Environmental Culture: The Ecological Crisis of Reason*. Routledge, 2002.

Pollan, Michael. *Second Nature: A Gardener's Education*. Grove Press, 1991.

Schleusener, Simon. "The Machine is the Garden: Concepts of Ecology and Nature in the Anthropocene." *The New Polis*, 8 Oct. 2019. Available from: https://www.thenewpolis.com/2019/10/08/the-machine-is-the-garden-concepts-of-ecology-and-nature-in-the-anthropocene/.

Searcy, David. *Ordinary Horror*. 2001. Reprint ed., Plume Books, 2002.

Simpson, Michael. "The Anthropocene as Colonial Discourse." *Environment and Planning D: Society and Space*, vol. 38, no. 1, 2020, pp. 53–71. https://doi.org/10.1177/0263775818764679.

Tsing, Anna Lowenhaupt. "Blasted Landscapes (and the Gentle Arts of Mushroom Picking)." *The Multispecies Salon*, edited by Eben Kirksey, Duke University Press, 2014, pp. 87–109.

Tsing, Anna Lowenhaupt. *The Mushroom at the End of the World: On the Possibility of Life in Capitalist Ruins*. Princeton University Press, 2015.

Vermeulen, Pieter. *Literature and the Anthropocene*. Routledge, 2020.

Wang, Dorothy J. *Thinking Its Presence: Form, Race, and Subjectivity in Contemporary Asian American Poetry*. Stanford University Press, 2014.

Williams, Raymond. "Ideas of Nature." 1972. *Problems in Materialism and Culture: Selected Essays*. Verso Books, 1980, pp. 67–85.

Williams, Raymond. "Nature." *Keywords: A Vocabulary of Culture and Society*. 1976. Reprint ed., Oxford University Press, 2015, pp. 164–169.

CHAPTER 10

"Relearning the Lessons of Land Reverence": Land Stewardship as Environmental Justice in the Writings of Leah Penniman and Natalie Baszile

Jan Beneš

Abstract

"[T]here is more to the story of the modern black farmer" than existing research suggests, argues Leslie Touzeau in her 2019 study. This chapter[1] contends that Leah Penniman's *Farming While Black* (2018) and *Black Earth Wisdom* (2023), along with Natalie Baszile's novel *Queen Sugar* (2014) represent ideal material for studying the modern Black farmer. The texts function as empowering, healing recordings and also uplifting fictional imaginings of generations of Black farmers and farming communities as well as of the ethos of Black agrarianism. Specifically, these texts theorize, articulate, and bear witness to how Black farmers of different generations and genders have (re) built their relationship to and found consolation in the land through stewardship. Additionally, by providing what Leah Penniman calls "dignified narratives of our relationship to the land" (Bittner), and by addressing African Americans' historical trauma of being enslaved, dispossessed, and disenfranchised, as well as by rewriting harmful stereotypes about working the land, the texts also perform environmental justice. The texts by Penniman and Baszile harness the reparative and healing qualities of literature and land as they address "black people's complex relationship to the land" (White 142) and interpret black farming and land stewardship as morally and economically uplifting, as resistance to and healing for the wounds of white supremacy and environmental racism.

Keywords

land stewardship – environmental justice – Black agrarianism – African American literature – *Farming While Black* – *Black Earth Wisdom* – *Queen Sugar*

---

1 This chapter is a result of the research project number 22–23300S "Environmental Justice in Ethnic American Literatures" financed by the Czech Science Foundation.

"Studies examining the experiences of young black farmers are virtually non-existent," writes Leslie Touzeau in her 2019 study, "'Being stewards of land is our legacy': Exploring the Lived Experiences of Young Black Farmers" (49). Her subjects are stewards of the(ir) land who farm "to take care of the land, pro-vide knowledge and resources to their communities, and maintain resilient links between their people, their history, and place." They seem to embrace the ideology of Black agrarianism and practice its central tenet that "land equals power in the black community." However, Touzeau notes, "further research into the experiences of young black farmers is needed and warranted [because] there is more to the story of the modern Black farmer" (58).

Cultural production such as literature may expand the story of modern Black farmers and the ethos they embrace. Since the artistic expression and intellectual thought of African Americans often builds on their lived experi-ence (Marable 1999), it seems logical to look for the stories of Black farmers in literature about and by Black farmers. This is especially necessary because, as Ian Finseth claims, "black labour has historically been hidden or marginal-ized in mainstream American representations of agricultural work" (230). At the same time, as Sterling A. Brown, an influential African American poet and literary critic in the first half of the 20th century, warns, one must beware of racially stereotypical plantation pastoral representations of Black farmworkers filled with false nostalgic depictions of social harmony (Anderson 86–87). It is perhaps these stereotypes as well as the painful history of dispossession why farm novels and farmers owning land rarely feature in modern African Ameri-can literature (cf. Wilson 2020; O'Donoghue 2020).

However, writings published in the last decade signal a shift. Farmer-activists like Leah Penniman and the novelist Natalie Baszile have written extensively about the past, present, and future of Black farming. Their texts serve as (text) books, farming and spiritual guides for restorative and uplifting engagements with nature and land, as well as fictional yet educational dramatizations of the lives, struggles, and values of Black farmers. They are not only inspired and driven by the history, legacy, and ethos of Black farmers, but they also aim to inspire and teach the present and future generations of Black farmers – they thus represent ideal material for further study of the modern Black farmers.

As this chapter argues, texts like Penniman's *Farming While Black: Soul Fire Farm's Practical Guide to Liberation on the Land* (2018) and *Black Earth Wisdom: Soulful Conversations with Black Environmentalists* (2023), along with Natalie Baszile's novel *Queen Sugar* (2014), function as empowering and healing por-trayals and, in the case of *Queen Sugar*, also fictional uplifting depictions of multiple generations of Black farmers as well as of the ideology and ethos of Black agrarianism. More specifically, these texts theorize, bear witness to, and teach about the ways in which Black farmers of different generations and genders have (re)built their relationship to and found consolation in the land through stewardship.

Finally, by providing what Penniman calls "dignified narratives of our relationship to the land" (Bittner), by addressing African Americans' historical trauma of being enslaved and dispossessed, and by rewriting stereotypes and misconceptions about working the land, the aforementioned texts perform environmental justice. In other words, the authors address environmental issues – such as environmental racism, dispossession, the disproportionate effects of climate change and environmental challenges like droughts, hurricanes, and extreme weather on Black (farming) communities – alongside social justice ones and represent them in literature. As Julie Sze explains, "culture matters in environmental and social struggle" and "literature offers a new way of looking at environmental justice, through visual images and metaphors, not solely through the prism of statistics." This new way of looking helps identify "the 'real' problems of communities struggling against environmental racism, and is simultaneously liberated from providing a strictly documentary account of the contemporary world." Analysing such cultural texts offers "an alternative strategy to analysing the roots of environmental racism" (76–77). As Sze proposes, literary texts may thus complement scientific studies like Touzeau's, and function as a literature of environmental justice.

This is precisely the type of analysis that this chapter engages in. The texts by Penniman and Baszile analysed here harness the reparative and healing qualities of literature and nature (Fiskio, *Climate Change* 35) as they address "black people's complex relationship to the land" (White 142). They depict journeys toward repairing this relationship with land burdened by the history of slavery, dispossession, (forced) migration, and institutional racism, while interpreting Black farming and land stewardship as morally and economically uplifting, as resistance to and healing for the wounds of white supremacy and environmental racism.

1   Dispossession, Black Agrarianism and Land Stewardship

Farming While Black, Black Earth Wisdom, and *Queen Sugar* promote and envision different variations of land stewardship by Black farmers. Land stewardship, defined as "the responsible use (including conservation) of natural resources in a way that takes full and balanced account of the interests of society, future generations, and other species, as well as of private needs, and accepts significant answerability to society" (Worrell and Appleby 1), represents an integral part of Black agrarianism. This ideology comprises "the agricultural, botanical, and culinary knowledge and practices brought to the United States by enslaved Africans, the knowledge produced and skills created by enslaved Africans and African Americans in the US" as well as the "knowledge and practices formed by African Americans since emancipation into the present" (Fiskio et al., "Cultivating Community" 2). Simultaneously, Black agrarianism may be defined as an ethos of "deep affirmative ties to the land [...] in the face of ongoing discrimination," which threatens to erase the aforementioned knowledge, practices, as well as any close relationship to the land (K. King et al. 691). However, it is also an ideology "rooted in the collective experiences of slavery, white supremacy, and systemic discrimination [which] not only advocates the virtues of hard work and self-sufficiency, but [...] is also a form of territorial liberation" (Touzeau 48). The building blocks of this ethos are "Black landownership and agrarian lifestyle," which are seen as "a means to escape the white-dominated system and affirm one's political and civil rights" (48). Black agrarianism, therefore, includes a strong social and environmental justice dimension.

Black agrarianism has been put to the test throughout American history, but its persistence indicates its significance for African Americans. In today's United States, Black farmers form a resilient, but small group. Only around 1.4 percent of American farmers, or about 45,000 farmers, are Black; they make up only 0.5 percent of total US farm sales ("Black Farmers in the US"). Although the situation is dire, it represents an improvement compared to 1982, when the U.S. Commission on Civil Rights declared that "the effects of historical discrimination and structural inequities could result in the extinction of black farms in this country [by the year 2000] if immediate measures are not taken to counter the biases presently built into the system" ("The Decline of Black Farming in America" 70). Yet, rather than dying out, Black farmers persevered – and sued. In 1999, the Supreme Court ruled in *Pigford v. Glickman* that the United States Department of Agriculture (USDA) had throughout the 20th century systematically discriminated against Black farmers as it repeatedly denied federal loans, made them wait longer for loan approval, and offered unfavorable loan terms to Black farmers. Consequently, many Black farmers faced foreclosures, debt, and lost their jobs as well as farms. At the time, the case was the largest civil rights settlement in US history, and around thirty thousand claimants received settlements of approximately $50,000 per farmer ("The Impact and Implications").

The USDA racist practices destroyed the fortunes of Black farmers and tested the viability of Black agrarianism in unprecedented ways. In 1920, as a result of what W. E. B. Du Bois described as African Americans' "land hunger" (537), due to the collective activities by organizations like Colored Farmers' Alliance or the UNIA (White 2018; Reid and Bennett 2012) and the financing and leadership by Booker T. Washington, there were as many as 926,000 Black farmers, which formed 14 percent of all American farmers. However, by 1997, the number had shrunk to only 20,000 (Gilbert, Sharp, and Felin 2) – since 1920, the number has decreased by 98 percent, a rate much higher than that of white farmers (Wood and Gilbert 43). Surprisingly, as Pete Daniel points out, "black farmers suffered the most debilitating discrimination during the civil rights era, when laws supposedly protected them from bias" (5). In fact, "Between 1940 and 1969, the rural transformation, fueled largely by machines and chemicals and directed by the USDA, pushed nearly 600,000 African Americans" off the land (7). In the 1960s alone, "185,000 black farmers left the land, and only 87,000 remained when Richard Nixon entered office" (7). Overall, Black farmers lost at least $326 billion worth of acreage during the 20th century (Francis 2).

Nevertheless, throughout the periods of enslavement, Jim Crow laws, the civil rights movement, and the post-civil rights era, Black farmers have developed and held onto the ideology of Black agrarianism, with its strong environmental justice dimension. As early as the 1840s, black and white abolitionists embraced agriculture as the most noble endeavor for the freedmen, for they believed that "the purity, virtue, and industry of farm life would restore the wounds of slavery and racism." Early Black agrarianism "championed farm life from an abolitionist and anti-racist position" – land ownership and farming were seen as reparative as they would "empower those who took to it and [...] it would secure African Americans against unrelenting racism and structural violence" (Feeley 309–310). Freedmen would own their land and would no longer work under a master's supervision, but instead "establish their own control over the natural world from a position of individual independence and political and social equality" (Smith, *African American Environmental Thought* 52).

This principal tenet of Black agrarian thinking was borrowed from Jeffersonian democratic agrarianism, which held that "owning a farm and cultivating it through one's own labor creates a character ideally suited to republican government" (Smith, *African American Environmental Thought* 57). According to Jefferson and fellow proponents of democratic agrarianism, explains Kimberly Smith, "agricultural labor cultivates virtues conducive to good citizenship, including self-sufficiency, industriousness, humility, spirituality [...] and prudence" (*African American Environmental Thought* 56). It stands to reason then, argued Black agrarian thinkers, that "the slaves who did the bulk of agricultural labor in the South had the best claim" to the right to the land and to citizenship. Any attempts to impede African Americans' efforts to acquire, own, and live on the land would alienate them from land, turn them into poor stewards of it, and create little incentive for them to be(come) honorable citizens. In this way, Black agrarianism transvalued agrarian labor, land ownership and stewardship from a "mark of ignominy" to "a source of pride" and "a strong basis on which to build [African Americans'] sense of self" (Smith, *African American Environmental Thought* 74).

As Du Bois put it, landlessness for African Americans in the post-Reconstruction period threatened to turn to homelessness. However, citizenship based on land ownership would allow African Americans to develop a sense of belonging as well as "an interest in and affection for the land that should lead to good stewardship" (Smith, *African American Environmental Thought* 52). In turn, this affection would help Black farmers develop "virtuousness" which would then "be seen in and as the health of the land" (Feeley 315). The freedom to own and farm land would help reform and shape the character of former slaves, which had been broken by slavery. Importantly, this recovery would manifest in the ecological improvement of the land.

The ecological concern for the reparative qualities of agrarian life stemmed from the reasoning by early Black agrarians that "race slavery and post-Emancipation racial oppression put black Americans into a conflicted relationship to the land – by coercing their labor, restricting their ability to own land, and impairing their ability to interpret the landscape" (Smith, *African American Environmental Thought* 18). In short, those who were forced to work the land were demeaned, no matter what the literature of the plantation tradition attempted for years to nostalgically argue (Brown). Moreover, the land, too, was "scarred" and demeaned by slavery. In fact, as Frederick Douglass and W. E. B. Du Bois repeatedly claimed, "America – not just the political community but the physical terrain – [was] a land cursed by injustice and in need of redemption" (Smith, *African American Environmental Thought* 18). In this way, Black agrarian thinkers made land stewardship central to their ideology. Instead of wondering "how to protect the natural world from human interference," they asked "how to facilitate responsible and morally beneficial interaction with nature" (Smith, *African American Environmental Thought* 19) – how to redeem the land and find dignity in working and stewarding it despite the effects of slavery and sharecropping.

Black agrarians have voiced these ecological concerns since the 1840s. In fact, Kimberly Smith claims that the environmental justice movement of today has "evolved over many decades of political activism [and Black environmental thought] aimed at rectifying blacks' relationship to the land" (*African American Environmental Thought* 16), with Du Bois serving as "an important forerunner" of the movement (Smith, "W. E. B. Du Bois" 223). Indeed, abolitionists believed that "slavery was the root of ecological decline" as the soil on large Southern plantations in the 1840s and 50s became exhausted (Feeley 309). The afore-mentioned scarring and degradation of American land, too, was both moral and ecological. The morally degrading effects of slavery – alongside its over-reliance on cheap labor – translated into poor stewardship. As Douglass and numerous other black and white abolitionists repeatedly argued in slave nar-ratives, slavery negatively affected both the black slaves and the white slave owners. This dynamic manifested in their complicated relationship to the land, too.

The social and environmental justice aspects of Black agrarianism became even more prominent in the 20th century. Du Bois, George Washington Carver or Booker T. Washington repeatedly highlighted "the value of agricultural work as a form of independence and self-reliance." In this way, explains Monica M. White, they "laid the foundation for an important counternarrative," inherent to Black agrarianism, which tied agriculture to freedom (61). Given the fact that millions of African Americans left Southern agriculture for the allure of the industrial metropolises during the Great Migration, often with the wish to leave agricultural labor behind, the Black agrarian narrative represented the appreciation of and encouragement for those who decided or managed to stay and work the little land they owned or were tenants on.

Multiple cooperative organizations were established throughout the 20th and 21st centuries to support Black farmers and promote the ethos of Black agrarianism. Monica M. White's *Freedom Farmers* (2018), for example, dives into the aims and activities of cooperative farmer groups like Fannie Lou Hamer's Freedom Farm Cooperative (FFC), North Bolivar County Farmers Cooperative (NBCFC) or the Federation of Southern Cooperatives (FSC). They, like the National Federation of Colored Farmers, an organization founded in the 1930s, serve as examples of efforts to establish sustainable, self-sufficient, cooperative, and politically engaged communities of black farmers/landowners, which functioned as resistance against exploitation and racial discrimination in places like Mississippi, North Carolina or Michigan.

National organizations such as Marcus Garvey's UNIA or the Nation of Islam (NOI) also emphasized land ownership and the agrarian life as a means of developing economic and political autonomy. The UNIA's 1920s version of Black agrarianism deemed work in agriculture "not just a practical means to achieve autonomy but a religious duty that God would reward" (Roll 136). For example, one 1925 editorial in *The Negro World* stated that "We believe in the farmer and in ownership in the soil as the most independent life." It advised the readers "to own land wherever they can and to raise their own home supplies, and thus become independent of the country storekeeper and the credit system" (Roll 137). As Jarod Roll claims, Garveyism, a Black nationalist movement, with its encouragement to work the land to find freedom "was strongest in communities anchored by significant numbers of landowners" – almost a third of UNIA's 1,176 divisions around the world were located in Southern farming communities, especially the cotton regions of "Arkansas, Mississippi, Missouri, and Oklahoma" (134). In 1945, the organization opened its own farm, Liberty Farm, in Ohio (White 17).

The NOI, whose often-used slogan reads "The farm is the engine of our national life," also turned to agriculture to provide healthy food for the urban communities in the 1960s. It established and owned the Salaam Agricultural Systems, with "approximately 13,000 acres throughout the South" (White 19; McCutcheon 65). Both Elijah Muhammad and Malcolm X often highlighted the revolutionary potential of owning land, producing the community's own food, and serving as the khalifah, or stewards of the land: "Revolution is based on land. Land is the basis of all independence. Land is the basis of freedom, justice, and equality," claimed Malcolm X in 1965 (4). In the 1990s, the organization took up agrarianism once again when it established Muhammad Farms with 1,556 acres of land (McCutcheon 61).

Black farmers also played a crucial role in the Christian-led civil rights movement. Indeed, the access to and ownership of land as a means of achieving freedom and self-sufficiency – the core concepts in Black agrarianism – was very much in tune with the emancipatory tenets of the movement. And it was Black farmers who led the movement in states like Mississippi, with Fannie Lou Hamer and other family farm owners providing support and organizational infrastructure for the Freedom Summer of 1964 (Harris 254). It was black farmers in South Carolina, too, who sued in the *Briggs v. Elliott* case, the first of the *Brown v. Board of Education* cluster of cases challenging racial segregation in education (254). Even Martin Luther King, Jr. acknowledged the importance of land for African Americans' struggle for civil rights and economic equality in a speech in 1968:

At the very same time that America refused to give the Negro any land, through an act of Congress our government was giving away millions of acres of land in the West and the Midwest – which meant that it was willing to undergird its white peasants from Europe with an economic floor. But not only did they give them land, they built land grant colleges with government money to teach them how to farm. Not only that, they provided county agents to further their expertise in farming. Not only that, they provided low interest rates in order that they could mechanize their farms. Not only that, today, many of these people are receiving millions of dollars in federal subsidies not to farm, and they're the very people telling the Black man that he ought to lift himself by his own bootstraps. (M. King, "The Other America")

King here laments the disinvestment from those African Americans who are trying to continue farming and stewarding the land despite ongoing discrimination. However, the disinvestment continues today as racial discrimination and disparities in farming also seep into how the 48,000 Black farmers steward their land in the face of climate change and environmental challenges like droughts, hurricanes, and extreme weather. Because of the emphasis by USDA on consolidating land ownership in large-scale operations, Black farmers, who typically own and run small farms, historically lack access to credit, loans, crop insurance coverage or federal subsidies necessary to address the challenges caused by the changing climate. They are therefore much more vulnerable to the effects of climate change (Furman et al.; Zhang). In other words, they have to deal with what Robert Bullard terms environmental racism – where communities of color are systemically positioned to be disproportionately burdened with pollution and other environmental problems (Bullard 23). In this case, Black farmers – if they want to own and steward their land – have to shoulder the expensive and devastating environmental challenges of climate change without having access to resources available to white farmers.

Farming and stewarding one's own land, therefore, remains a revolutionary act for African Americans. It shows resilience of both the farmers – entire generations of them – but also of the ideology of Black agrarianism, with its social and environmental dimensions. If land stewardship involves the management of natural resources, while taking account of the interests of society and future generations, then Black agrarian history, theory, and practice show that Black farmers have always stressed the importance of owning and farming land sustainably in order to provide for their communities and future generations. Land has always been seen as a gateway to both citizenship and the future; a means of gaining civil rights and ensuring the provision of resources for the communities and families of farmers. At the same time, Black agrarianism has helped retain and celebrate the values of self-sufficiency and hard work among African Americans. Despite centuries of being forced to work the land, being dispossessed and driven off the land, being robbed of and having their agricultural, botanical, and culinary knowledge and practices erased, Black farmers have retained deep affirmative connections to the land.

2   Leah Penniman: Teaching Black Land Stewardship and Agrarianism

During the Covid-19 pandemic and in the aftermath of the Black Lives Matter protests in the summer of 2020, Leah Penniman's *Farming While Black: Soul Fire Farm's Practical Guide to Liberation on the Land* (*FWB*) gained widespread popularity. The book's combination of Black farming practices and agrarian solutions to the hotly-debated issues of medical and environmental racism, housing, and policing resonated with the moment's *zeitgeist*. Penniman gave interviews to *Vogue*, spoke at Harvard, and made appearances on talk shows. Hers was both a vision and direction for Black farming as she became the voice and face of modern Black agrarianism (Reese and Cooper 47), following in the footsteps of farmer-activists like Fannie Lou Hamer, who used agriculture "as a strategy of resistance, resilience, and community building" (White 62).

Penniman recently complemented *Farming While Black* with *Black Earth Wisdom: Soulful Conversations with Black Environmentalists* (*BEW*). Together, they represent a mixture of practical, historical, and spiritual introductions to Black, environmentally just land stewardship within the framework of Black agrarianism. The texts have an autobiographical bent in the form of traditional nature writing, but include interviews, testimonials, and poems as well. Penniman does not only share her know-how of urban farming, but also roots her texts and expertise in the wisdom and knowledge of previous generations of Black farmers. In essence, she educates her readers on the agricultural, botanical, and culinary knowledge and practices of previous as well as current generations of Black farmers and combines it with a celebration of farm work.

Whereas most of *FWB* is written and structured in the form of a practical guide on finding land, budgeting, seeds, crops, cooking as well as trauma-healing and interracial-community building, the opening section tells Penniman's life story and outlines her Black agrarianism. The chapter "Black Land Matters" immediately reveals its theme as Penniman explains that she "found it very difficult to understand who I was" as a biracial child. Bullied and jeered at, she would run into the woods to find escape from her white schoolmates and solace in the land. Eventually, while farming at a summer "Food Programme" camp, Penniman finds "home" working "shoulder-to-shoulder with my peers of all hues, feet planted firmly in the earth, stewarding life-giving crops for Black community." In short, she realizes that Black land matters (1).

And yet, Penniman also shares her struggles with the sensation of double consciousness, as she farms and attends agricultural conferences. Because the field is so "white-dominated," she believes that her "ancestors who fought and died to break away from the land would roll over in their graves to see me stooping. I struggled with the feeling that a life on land would be a betrayal of my people" (*FWB* 1–2). The stigma of being a Black farmer haunts her. However, as Penniman explains in both texts, she gradually learns – through networking with other Black farmers and studying the history of Black agrarianism – that even though there is truth in seeing the American land as "the scene of the crime" (*FWB* 19), it "was never the criminal" (*BEW* 146). She therefore rejects the stigma of farming while Black.

It is this gathering of information that Penniman bases her contribution to Black agrarianism on and highlights as the key to destigmatizing and recovering one's relationship with the land. As she argues in *BEW*, "Part of the work of healing our relationship with soil is unearthing and relearning the lessons of land reverence and agrarian innovation from the past. Part of healing racism in America is to give credit where it is due to Black, Indigenous, and Asian peoples for their contributions to regenerative ways of tending agroecosystems" (146).

Penniman's usage of the term "relearning" is of particular significance here – like Toni Morrison's concept of rememory in her 1987 novel *Beloved*, Penniman's relearning implies a nexus of personal experience, collective memory, and cultural knowledge (N. King 150). It implies wisdom forgotten or suppressed, yet revived through creating affirmative ties to one's ancestors and absorbing their knowledge. In *Beloved*, Denver may only save her mother's life by learning about slavery and the scars it has left on her mother's body, but for that to happen she must re-join the Black community in Cincinnati, Ohio, from which her mother has been isolated for many years. Penniman, too, has learned that she has been isolated from the wisdom of her community and the healing and uplift it may offer, and therefore seeks the guidance of elder Black farmers to see "how miseducated" she is (*FWB* 3).

Land stewardship represents a crucial part of this relearning process. As Penniman emphasizes, "to learn of our true and noble history as farmers and ecological stewards is deeply healing" (*FWB* 29). In the "Soil" section of *BEW*, which deals with land loss, Penniman records a discussion with Savi Horne from the North Carolina Association of Black Lawyers Land Loss Prevention Project and the National Environmental Justice Advisory Council of the US Environmental Protection Agency. Horne explains that "being good stewards of the planet [...] means pouring libations to recognize that we stand on the shoulders of our agrarian ancestors. [...] When we ground ourselves in an African agro-ecological framework, it helps us fight for economic, gender, and planetary justice" (144). Whereas the forgetting of the rich agrarian history and legacy of Africans and African Americans is a direct result of land dispossession and environmental racism, relearning how one's ancestors stewarded the land is part of and leads to environmental justice.

In this way, Penniman's texts may be read as educational literature of environmental justice. The farmer-writer frames them as textbooks – Penniman has been working as a high school biology and environmental science teacher for almost two decades and her texts are imbued with a multitude of relearning moments. Quoting Toni Morrison's *Song of Solomon* (1977), Penniman explains that she feels compelled to "pass it on" that "Our Black ancestors and contemporaries have always been leaders in sustainable agriculture and food justice movements [...] It is time for us all to listen" (*FWB* 40). That is why alongside the practical information on setting up a community farm or garden, almost every chapter of *FWB* includes a short "Uplift" section, through which readers may relearn about *Pigford v Glickman*, Hamer's Freedom Farm Cooperative, Food Hubs of the Black American South or Indigenous soil testing. The book thus aims to create a sense of rich, albeit little-known agricultural history and legacy. Often combined with detailed step-by-step guides on turning a piece of land into a farm or garden, *FWB* encourages prospective Black farmers and fills them with pride. In short, Penniman's teaching erases stigma from land stewardship and labor.

However, while Penniman celebrates relearning from her ancestors, she also devotes parts of her opening chapter in *FWB* to the relearning process among the Black youth. Explaining how Black lives and Black land are tied together, she highlights that Black agrarianism leads to liberation. Penniman tells the story of how her farm began offering farm work as a form of community service "in lieu of punitive sentencing" (34). She explains that it is relearning one's "connection to the land" that helps those stuck in the school-to-prison pipeline to regain "full respect of their humanity." When one of "the most hardened and defended child[ren]" visiting the Soul Fire Farm feels "his grandmothers reach[ing] for him through the earth under his feet [on the garlic field] and remind[ing] him that there is a point," he begins to "weep." Whereas Penniman describes how full of fatalism – "a form of internalized racism" – this young man is at first, he ultimately finds purpose on the land, she says (33–34). Penniman thus reaches across generations to multiply the effect of relearning through reconnecting with the land.

Penniman also extends her relearning across races. She reminds her readers that "farm management is among the whitest professions," whereas "farm labor is predominantly Brown and exploited" (*FWB* 32). That is why she highlights the Black Latinx Farmers Immersion programme "as a humble attempt to rewrite part of this story [...] of Blacks, Latinx, and Indigenous people working in the food systems [being] more likely than whites to earn lower wages, receive fewer benefits, and live without access to healthy food" (37). Penniman's relearning thus involves rewriting – by celebrating farming and land stewardship as noble and moral endeavors with rich history, as well as by providing hands-on guidance on how to achieve environmental and food justice.

Finally, *BEW* adds a spiritual level to Penniman's relearning efforts. It seeks answers to how to best environmentally steward the land by relating discussions on Muslim and Christian practices and approaches. After all, Christian and Muslim organizations have always shaped and participated in Black agrarianism. Most importantly, however, Penniman – with a Unitarian Universalist background – shows that environmentally-conscious land stewardship connects religions and therefore people. She teaches her readers through a series of interviews with religious leaders, philosophers, and activists that Abrahamic faiths are driven by an "environmental imperative" (68) and that there are clear connections between faith and environmental stewardship. She also reminds the readers that "Black and Hispanic Christians are more concerned about climate change than both white Christians and secular people" and that "Black and Hispanic clergy are also more likely than white clergy to talk about climate change with their congregations" (68). Alongside practical guidance – complemented with her own testimonials and agri-centered stories of uplift – Penniman's educational texts thus offer spiritual guidance and arguments for stewarding the land with pride.

Overall, a holistic image and expression of contemporary Black agrarianism emerges through Penniman's literary efforts to help her readers relearn how to heal from American racism and rediscover dignity in owning and stewarding land. Penniman repeatedly implies that farm work is hard work, but that, too, is integral to the Black agrarian tradition. In fact, it is the willingness to own, steward, and work the land that makes Black agrarianism an ideology of resistance and self-sufficiency. As Penniman puts it, "to farm while Black," despite the long history of dispossession, environmental racism, and stigmatization of farm labor, "is an act of defiance against white supremacy" (*FWB* 39). Consequently, as her texts repeatedly articulate and advocate, "Owning our own land, growing our own food, educating our own youth, participating in our own healthcare and justice systems – this is the real source of power and dignity" (*FWB* 37).

3   *Queen Sugar*: Learning Black Land Stewardship and Agrarianism

"Artists and poets are essential because they connect us to what we cannot see. They open our eyes to what we are blind to [...]. Poets who love the earth and express it in a way that you can feel it make you more likely to experience that connection yourself," Alice Walker, author and activist, tells Leah Penniman in one of the final chapters of *BEW* (252). Penniman conceptualizes her two books as "love songs for the land and her people" ("Leah Penniman") and her nature writing helps readers relearn the connection that Walker speaks of. In her novel *Queen Sugar*, Natalie Baszile also assumes the role of a poet uncovering that which we rarely see: the modern Black farmer. *Queen Sugar*'s story of Black agrarianism, too, is based on the process of (painful) relearning, which, however, brings uplifting dignity, healing, and the promise of a future to her Black farmer.

The novel follows the transformative journey of California-based Charley Bordelon, who unexpectedly inherits "eight hundred acres of sugarcane land" in Louisiana from her father, from a lost "city girl" (Baszile 4) to an "honest-to-goodness cane farmer" (156). Throughout the novel, Charley navigates the overwhelming project of relearning how to connect with and steward a land and a crop burdened with the violent history of chattel slavery, sharecropping, as well as dispossession, while she juggles responsibilities toward and demands from her daughter as well as her Louisiana family.

When Charley becomes "the lucky owner of LeJeune's plantation" (Baszile 49), she is just that: an owner. She knows nothing about farming and stewarding land. She has "never seen real sugarcane before yesterday" (3) and does not feel like "a country girl. Not even a little" (4). She does not understand why her father left her a sugarcane farm "in south *bumfuck* Louisiana" (8). Charley's lack of knowledge is only compounded when she finds out that instead of idyllic "fields so splendidly verdant she'd feel short of breath just looking at them" (9), her farm is covered with "stunted cane stalks" with "straggly leaves a starved shade of pale green" (9). By the end of the opening chapter, Charley is despondent, feeling that "she was a girl who kept losing things: she lost her husband in a holdup [and] she almost lost her daughter. She lost her father to cancer, and now she was about to lose his strange and unexplained legacy, this sugarcane farm" (14). As Charley stands looking at what she imagined as a promised land, she realizes – somewhat similarly to Penniman early in her career – that although the land belongs to her, she may not belong to the land.

The semantics of how Baszile narrates Charley's relearning of Black agrarianism are crucial here. Indeed, her father's legacy stands for the land that Charley inherits. However, it is later revealed to her that the legacy represents her father's espousal of Black agrarianism, too. Although Charley remembers "how much he hated the South" (123), she learns from her grandmother that, at the age of thirteen, Ernest "got up before dawn and walked four miles" to get himself "hired onto one of the crews" of cane cutters. When he attempted to drink from a bucket for the field hands, "something caught him upside his face. Said it felt like a hunk of metal." Ernest got hit in the face with a shovel and told to "let the white drink first," because the white sugar cane farmer "couldn't stand seeing a black boy drink ahead of him." Charley's grandmother admits that Ernest did not tell her about the incident until he finished the job; she also was not aware that he had bought the entire farm until Charley arrived. However, "I can't tell him how proud I am," she admits to Charley (124–25). Owning farming land is here portrayed as a form of defiance (McInnis 768) and land justice, and the Black-owned land becomes a counter-site (Niewiadomska-Flis 174).

Ernest's perseverance to fulfil his dream of eventually owning the land where white supremacists humiliated him is also presented as a path to follow for Charley. Part of the second wave of the Great Migration, Ernest becomes a multiple-property owner in California, telling Charley that "real estate was the only thing worth buying" (30). By leaving her a sugarcane farm instead of rental property, however, Ernest leads Charley toward relearning what it means to own productive land; she will relearn about her roots, her Louisiana family, and his own Black agrarian dream. She will become one with the land. By also leaving her a statue of The Cane Cutter, Ernest teaches Charley – like Penniman strives to do in her writing – that there is a strong agrarian legacy that can be passed on within the family and the community.

Although it takes some convincing, Charley eventually finds the right person to help her relearn how to steward her land: Prosper Denton. As his first name implies, Charley will find prosperity with him, but she first needs to relearn her lessons. In fact, the novel is built on relearning scenes and the text reads like a dramatization of Penniman's guide for beginner farmers. Denton begins teaching Charley with what Jarvis C. McInnis calls the juxtaposition of "the political economy of the contemporary sugar industry with the history of black land ownership and farming" (767). "This ain't no game," he explains to Charley, "You're young, you're not from around here, you've never worked cane, and frankly, you being a woman's gonna work against you. [...] And colored on top of it?" (Baszile 32). Charley gradually finds out what Denton means when she is referred to as "a black surfer chick" (160) and told that "some people prefer the old ways" (50). As McInnis argues, the racism and sexism make Charley's character development a strong "critique of the plantation's history of racial and sexual violence and economic exploitation" (770).

After Charley reassures Denton that she is "not afraid of hard work" (90), he begins teaching her land stewardship. Sugar cane farming terminology, the machinery needed, the season schedules, budgeting – Charley eventually grasps all of these. The most eye-opening relearning moment, however, comes when Denton asks Charley to physically connect with the land, to become one with it. While Charley notes down everything Denton says, he explains to her, "All that scribbling won't do you any good if you don't let this get inside you."

Charley then tastes the soil she owns, but does not "taste anything." The second time around, however, an epiphany arrives:

> She sniffed: wood smoke, grass, damp like a sidewalk after it rained. She tasted: grit, fine as ground glass, chocolate and what? Maybe ash? She closed her eyes as soil dissolved over her tongue, and slowly, slowly, almost like a good wine, the soil began to tell its story. She tasted the muck, and the peat, and the years of composted leaves, the branches and vines that had been recently plowed under, and the faint sweetness the cane left behind. She swallowed: a moldy aftertaste she knew would stay on her tongue for the rest of the afternoon. And though she didn't yet know the terms to describe what she had experienced, she understood a little more clearly what Denton was trying to teach her. (117–118)

Although this scene is merely the beginning of Charley's journey to becoming a full-fledged Black farmer, it does illustrate the tenets of land stewardship within the framework of Black agrarianism. It depicts how Charley creates affirmative ties to the land: she no longer perceives soil as dirt; the land loses its stigma. Furthermore, Charley is relearning the agricultural knowledge, practices, and skills of her ancestors. And she is relearning them from a man who has never owned much farm land, yet his expertise as a land steward is such that "though he was the only black farmer on the lot [...] the younger farmers addressed him with respect, even admiration, while the older men greeted him like a brother" (Baszile 158). It is symbolic that Charley decides to honor Denton by agreeing "on a sixty-forty split" (192) of profits with him. Representing a father figure in the novel, Denton is similar to Ernest in his espousal of Black agrarianism – by making Denton "a partner" (160) in the business, Charley honors her father as well. Moreover, sharing profits and the land with another Black steward represents land liberation, so coveted among Black agrarians.

Furthermore, as Charley relearns under Denton's tutelage, she also embraces the most stigmatized part of being a Black land owner and steward: hard work. Whereas Charley's "disinherited brother" (Baszile 205), Ralph Angel, believes that "old manual labor [...] doesn't build anything but an aching back" (199), she partakes in the backbreaking work on a sugarcane plantation. During the planting season, "without a thought, Charley ran out to the field [and] joined the crew, pulling armloads of cane stalks off the back of the wagon." Although "it felt to her that a tsunami of cane was coming at her [...] there was no stopping." One of the hired laborers observes her with "a mixture of amusement and admiration. Charley imagined what he'd tell his buddies when he met them for a beer after work: that he was working for a crazy black woman from

California who not only owned the land, but got behind the wagon and planted cane herself." At the end of the shift, another worker tells her admiringly, "You work hard" (237–239). This constitutes a brief, but crucial sign of respect. It also represents a moment of pride for Charley, who relearns what it takes to own and steward her land.

The final lesson Charley relearns from Denton is that of treating the local African Americans and immigrant workers with respect. She pays them more than the white farmers, provides visas, healthcare, and other assistance. This means that she has good workers, who will return, thereby ensuring sustainability. Moreover, treating and paying them well functions as an element of environmental justice: migrant workers' lives and contributions are appreciated, the land is stewarded sustainably, and the effects of racial capitalism mitigated. Sugar cane cultivation remains hard work, but the Black landowner does not exploit the workers and even joins in the labor. The novel thus imagines environmentally just ethics of land stewardship.

In the end, however, no matter the lessons Charley has relearnt, she is on the brink of losing the farm. It is cooperative economics, a staple of Black agrarianism highlighted in Penniman's texts, that rescue her. Denton brings in a struggling white farmer to become a third partner, and Charley finds financial support among her Black community. Indeed, "it is men who come to her rescue" (Niewiadomska-Flis 180), but the novel also depicts an integral role of Charley's female relatives as a supportive cog in the "gendered economies of south Louisiana's sugar industry" (McInnis 769). At the end of the novel, now that her farm has been saved and she can imagine a future, Charley "looked through the window into the powder-blue sky and imagined her fields, the rows of cane – her cane, her father's cane – looking lush and orderly like the fields she passed when they drove in all those months ago" (Baszile 366). Despite the systemically racist and sexist obstacles, Charley has become a modern Black woman farmer.

4   The Modern Black Farmer and Steward

In a speech to Tennessee farmers in 1873, Frederick Douglass called agrarian life "a refuge for the oppressed. The grand old earth has no prejudices against race, color, or previous condition of servitude, but flings open her ample breast to all who will come to her for succor and relief" (Douglass, "Agriculture and Black Progress"). As literary expressions of modern Black agrarianism, Leah Penniman's and Natalie Baszile's texts echo and dramatize Douglass' statement in their depictions and dramatizations of the varieties of Black land stewardship – the consolation and comfort one may derive from intimate contact and relationship with the land. Their Black farmers are male and female, they strive to heal the wounds of land dispossession, environmental racism, and the stigma of working the land. The texts offer models to emulate and celebrate the legacy of Black agrarianism. They also propose the ethics of environmentally just and dignified land stewardship. They offer relearning, healing, and uplift.

Black farming represents, as Jay Fiskio puts it, "both a space of imagining a different world and a real place to grow freedom" (*Climate* 8). *Farming While Black*, *Black Earth Wisdom* and *Queen Sugar* allow for and imagine precisely this freedom in their function as literature of environmental justice. The texts mix social justice with environmental concerns and give voice to and guidance for those who have been historically dispossessed and exposed to environmental racism. They also provide real and fictionalized blueprints for the journey of a modern Black farmer from environmental discrimination to stewarding and finding healing, moral and economic uplift, community, and justice on the land.

Bibliography

Anderson, David R. "Sterling Brown and the Georgic Tradition in African-American Literature." *Green Letters: Studies in Ecocriticism*, vol. 20, no. 1, 2016, pp. 86–96.

Baszile, Natalie. *Queen Sugar*. Penguin, 2014.

Bittner, Mark. "Farming While Black: 'People Are Tired of Armchair Activism.'" *Medium. com*. https://heated.medium.com/farming-while-black-people-are-tired-of -armchair-activism-748e0f5111c. Accessed 19 February 2024.

"Black Farmers in the US: The Opportunity for Addressing Racial Disparities in Farming." *McKinsey and Company*, 10 Nov. 2021. https://mckinsey.com/industries/agriculture /our-insights/black-farmers-in-the-us-the-opportunity-for-addressing-racial -disparities-in-farming. Accessed 19 February 2024.

Brown, Sterling A. *The Negro in American Fiction*. Kennikat Press, 1938.

Bullard, Robert D. "The Threat of Environmental Racism." *Natural Resources & Environment*, vol. 7, no. 3, 1993, pp. 23–26, 55–56.

Daniel, Pete. *Dispossession: Discrimination against African American Farmers in the Age of Civil Rights*. University of North Carolina Press, 2013.

Douglass, Frederick. "Agriculture and Black Progress: An Address Delivered in Nashville, Tennessee, on September 18, 1873." *The Frederick Douglass Papers*. https:// frederickdouglasspapersproject.com/s/digitaledition/item/17769. Accessed 22 May 2024.

Du Bois, W. E. B. *Black Reconstruction in America*. Harcourt, 1935.

Feeley, Lynne. "The Elevationists: Gerrit Smith, Black Agrarianism, and Land Reform in 1840s New York." *Environmental History*, vol. 24, 2019, pp. 307–326.

Finseth, Ian Frederick. *Shades of Green: Visions of Nature in the Literature of American Slavery, 1770–1860*. University of Georgia Press, 2009.

Fiskio, Janet. *Climate Change, Literature, and Environmental Justice: Poetics of Dissent and Repair*. Cambridge University Press, 2021.

Fiskio, Janet, Md Rumi Shammin, and Vel Scott. "Cultivating Community: Black Agrarianism in Cleveland, Ohio." *Gastronomica*, vol. 16, no. 2, 2016, pp. 18–30.

Francis, Dania V., et al. "Black Land Loss: 1920–1997." *AEA Papers and Proceedings*, vol. 112, 2022, pp. 38–42.

Furman C., et al. "Social Justice in Climate Services: Engaging African American Farmers in the American South." *Climate Risk Management*, vol. 2, 2014, pp. 11–25.

Gilbert, Jess, Gwen Sharp, and M. Sindy Felin. "The Loss and Persistence of Black-Owned Farms and Farmland: A Review of the Research Literature and Its Implications." *Southern Rural Sociology*, vol. 18, no. 2, 2002, pp. 1–30.

Harris, Carmen V. "You're just like mules, you don't know your own strength: Rural South Carolina Blacks and the Emergence of the Civil Rights Struggle." *Beyond Forty Acres and a Mule: African American Landowning Families since Reconstruction*, edited by Debra A. Reid and Evan P. Bennett, University Press of Florida, 2012, pp. 254–270.

"The Impact and Implications of the Pigford Settlement for Black Farmers." *Institute for Economic and Racial Equality*. https://heller.brandeis.edu/iere/news/press-releases /2022-01-18-pigford-project.html. Accessed 19 February 2024.

King, Katrina Quisumbing, et al. "Black Agrarianism: The Significance of African American Landownership in the Rural South." *Rural Sociology*, vol. 83, no. 3, 2018, pp. 677–699.

King, Martin Luther, Jr. "The Other America." *Grosse Point Historical Society*. https:// www.gphistorical.org/mlk/mlkspeech/. Accessed 25 April 2024.

King, Nicola. *Memory, Narrative, Identity*. Edinburgh University Press, 2000.

"Leah Penniman." *Soul Fire Farm*. https://www.soulfirefarm.org/leah-penniman/. Accessed 22 May 2024.

Malcolm, X. *Malcolm X Speaks*. Pathfinder Press, 1965.

Marable, Manning. *How Capitalism Underdeveloped America*. South End Press, 1999.

McCutcheon, Priscilla. "'Returning Home to Our Rightful Place': The Nation of Islam and Muhammad Farms." *Geoforum*, vol. 49, 2013, pp. 61–70.

McInnis, Jarvis C. "Black Women's Geographies and the Afterlives of the Sugar Plantation." *American Literary History*, vol. 31, no. 4, 2019, pp. 741–774.

Niewiadomska-Flis, Urszula. "Black Land Matters: Geographies of Race and Politics of Land in Natalie Baszile's *Queen Sugar*." *Pathologizing Black Bodies: The Legacy of*

Plantation Slavery, edited by Constante González Groba, Ewa Barbara Luczak and Urszula Niewiadomska-Flis, Routledge, 2023, pp. 164–190.

O'Donoghue, James. *"From Behind the Plow": Agrarianism and Racial Uplift in African American Literature, 1881–1917.* 2020. Southern Illinois University, PhD dissertation.

Penniman, Leah. *Black Earth Wisdom: Soulful Conversations with Black Environmentalists.* Amistad, 2023.

Penniman, Leah. *Farming While Black: Soul Fire Farm's Practical Guide to Liberation on the Land.* Chelsea Green Publishing, 2018.

Reese, Ashanté M., and Dara Cooper. "Making Spaces like Freedom: Black Feminist Praxis in the Re/Imagining of a Just Food System." *ACME: An International Journal for Critical Geographies* vol. 20, no. 4, 2021, pp. 450–459. https://doi.org/10.14288/acme.v20i4.2024.

Reid, Debra A., and Evan P. Bennett. *Beyond Forty Acres and a Mule: African American Landowning Families since Reconstruction.* University Press of Florida, 2012.

Roll, Jarod. "'The Lazarus of American Farmers': The Politics of Black Agrarianism in the Jim Crow South, 1921–1938." *Beyond Forty Acres and a Mule: African American Landowning Families since Reconstruction*, edited by Debra A. Reid and Evan P. Bennett, University Press of Florida, 2012, pp. 132–152.

Smith, Kimberly K. *African American Environmental Thought: Foundations.* University Press of Kansas, 2007.

Smith, Kimberly K. "W. E. B. Du Bois: Racial Inequality and Alienation from Nature." *Engaging Nature Environmentalism and the Political Theory Canon*, edited by Peter F. Cannavò and Joseph H. Lane, Jr., MIT Press, 2014, pp. 223–238.

Sze, Julie. "From Environmental Justice Literature to the Literature of Environmental Justice." *The Environmental Justice Reader*, edited by Joni Adamson, Mei Mei Evans and Rachel Stein, University of Arizona Press, 2002, pp. 1963–1980.

Touzeau, Leslie. "'Being stewards of land is our legacy': Exploring the Lived Experiences of Young Black Farmers." *Journal of Agriculture, Food Systems, and Community Development*, vol. 8, no. 4, 2019, pp. 46–60.

United States, Department of Agriculture, Commission on Civil Rights. "The Decline of Black Farming in America." Government Printing Office, 1982.

White, Monica M. *Freedom Farmers: Agricultural Resistance and the Black Freedom Movement.* University of North Carolina Press, 2018.

Wilson, Laura. *On Southern Soil: The Art and Ecology of Racial Uplift, 1895–1950.* 2020. University of Mississippi, PhD dissertation.

Wood, Spencer D., and Jess Gilbert. "Returning African American Farmers to the Land: Recent Trends and a Policy Rationale." *The Review of Black Political Economy*, vol. 27, no. 4, 2000, pp. 43–64.

Worrell, Richard, and Michael C. Appleby. "Stewardship of Natural Resources: Definition, Ethical and Practical Aspects." *Journal of Agricultural and Environmental Ethics*, vol. 12, 2000, pp. 263–277.

Zhang, Huanjia. "For Black Farmers, Climate Change Magnifies Existing Inequality." *Scienceline*. https://scienceline.org/2021/01/for-black-farmers-climate-change-mag nifies-existing-inequality/. Accessed 19 February 2024.

CHAPTER 11

Crimes, Wolves and Consolation: National Parks in Contemporary Crime Fiction

Šárka Bubíková

Abstract

This chapter argues that contemporary crime fiction can effectively address environmental issues and reflect on the human–nature relationship. It has the ability to address the complexities involved in protecting wildlife while still retaining the appeal of a popular genre. Using Nevada Barr's *Winter Study* (2008) and Charlotte McConaghy's *Once There Were Wolves* (2021), the chapter demonstrates how meaningful deliberations on nature and human attitudes toward it were enabled by several changes in the genre's formula. These changes include approaching the setting as a complex environment, as defined in Lawrence Buell's seminal work *The Environmental Imagination* (1995). Similarly to the new nature writing characterized by Robert Macfarlane (2013), the novels mix scientific and poetic elements by inserting scientifically accurate descriptions of wildlife and natural processes into their crime plots. The chapter further analyzes how the novels use national parks as settings to engage in a discourse on wilderness therapy, the consolation of nature, wildness, and nature conservation. Finally, it discusses the incorporation of wolves into the crime narrative, highlighting the contrast between the wolf's symbolic value and realistic depiction. Animal predation is sharply distinguished from human violence, specifically domestic violence, emphasizing the novels' environmental advocacy. Exonerating the wrongly suspected predators adds another level of consolation for the environmentally-minded reader.

Keywords

crime fiction – nature – national park – wilderness therapy – Nevada Barr – Charlotte McConaghy – wolf – domestic violence

1   Introduction

Nature writing of the non-fiction kind has been hailed as a particularly suitable vehicle for exploring human relationships to nature and wildlife, of the experience of nature as a place of refuge and (moral) regeneration. Yet, one can wonder if such issues can be meaningfully interwoven into what Jo Lindsay Walton and Samantha Walton (2018) describe as "ostensibly anthropocentric plotting" (3) of the crime fiction genre. Action-packed, plot-driven with a strong emphasis on rational deduction, crime fiction does not seem a very suitable vehicle for meditations on nature, wildlife, and human relationships to it, even if we admit, with critics like Marta Puxan-Oliva (2020), that the genre's conventions can provide "a unique tool [...] for depicting and discussing ecological crises and abuses, [as well as] for directly exposing the criminal acts they involve and their violent effects on people and the environment" (362).

The most obvious way for crime fiction to participate in the current environmental discussions is by centering the investigation on environmental crimes. These can include asbestos exposure (for example in Laura Lippman's *Baltimore Blues*), illegal disposal of hazardous waste (in Nevada Barr's *Ill Wind*, in Charles LoPinto and Lidia Llamas's *Countdown in Alaska*), oil spills (Dana Stabenow's *A Cold Blooded Business*), bribing officials to circumvent environmental protection laws (Barr's *Firestorm*), wildlife mutilation (Janet Dawson's *Don't Turn Your Back on the Ocean*), abuse of protected areas for private purposes (Elly Griffiths's *The Crossing Places*), poaching (in Barr's *Track of the Cat*, in C. J. Box's *Open Season*), etc.

Apart from investigating environmental crimes, I argue that the genre is capable of more subtle reflections on human–nature connections. In studying crime fiction from the environmental perspective, I follow Timothy Clark's lead:

> For an environmental critic, every account of a natural, semi-natural or urban landscape must represent an implicit re-engagement with what 'nature' means or could mean, with the complex power and inheritance of this term and with its various implicit projections what of human identity is in relation to the non-human, with ideas of the wild, of nature as refuge or nature as a resource, nature as the space of the outcast, of sin and perversity, nature as a space of metamorphosis or redemption. (Clark 6)

Traditionally, crime fiction did not provide much material regarding accounts of natural or semi-natural landscapes, even if brief remarks and more or less implicit attitudes to the location of crime were always present. If one wants to read these from an ecocritical perspective, one must, as Samantha Walton (2018) explains, refocus "to features of a text often dismissed as backdrops to human activity" (Walton 115).[1] Contemporary crime fiction, however, provides far richer material for ecocritical study as also documented by the recent publication of the *Routledge Handbook of Crime Fiction and Ecology* (2024).[2] Several changes or shifts traceable in crime fiction in the last decades have fostered this development: firstly, a shift from the predominantly urban focus; secondly, a shift in the approach to the place of crime; and thirdly, the introduction of a new type of investigator. The selected works by Nevada Barr and Charlotte McConaghy document these shifts very well. Moving away from the traditional urban milieu, they are set in a national park and employ an environmental approach to their setting. They also feature non-traditional investigators.

Nevada Barr situates her Anna Pigeon crime series[3] (since 1993, the so far latest nineteenth novel was published in 2016) in various American national parks. Although her protagonist is part of law enforcement, the novelty is that she is a park ranger whose official responsibilities include protecting people who visit national parks as well as the parks' natural resources, ecosystems, and wildlife.[4] Due to the specific setting, she often operates alone rather than in a team of other professional law enforcement officers and shows a character affinity with the private eye protagonist. Thus, the series can be arguably considered both a fresh modification of the police procedural, as Hans Bertens and Theo D'haen claim (77), or a version of the hard-boiled subgenre, as Karin Molander Danielsson suggests in her insightful analysis of Barr's 2001 novel *Blood Lure* (Danielsson 105). She does so with respect to the protagonist's characteristics and her mode of operation. My analysis refers to several volumes of the series, but the most attention is paid to *Winter Study* (2008, the

---

1   In the study, Samantha Walton provides an inspiring ecocritical reading of one of the genre's classics, Arthur Conan Doyle's *The Hound of the Baskervilles* (1902).

2   Interestingly enough, none of the novels discussed here is analyzed in the collection. Ruth Hawthorn mentions Nevada Barr's *Burn* in her insightful chapter on environmental injustice because *Burn* takes place in post-Katrina New Orleans; however, this novel is one of the few in the series not set in a national park and not concerned with environmental issues – its central theme is child trafficking.

3   The series has received significant critical acclaim. Its opening volume garnered both the Agatha Award and the Anthony Award for Best First Novel, and several other volumes were nominated for the Anthony Award. Apart from that, *Blind Descent* (1998) won both the Dilys and the Macavity Awards, and *Deep South* (2000) won the Barry Award. Nevada Barr received the Robin W. Winks Award for Enhancing Public Understanding of National Parks in 2011.

4   For more details about the park ranger's job, see, for example, the educational webpage "What Is a Park Ranger" (parkrangeredu.org).

fourteenth novel in the series), set in the Isle Royal National Park, a remote freshwater archipelago in Lake Superior, where Anna Pigeon arrives in January to join a wolf study team while the park is closed to the public for the winter. What should have been a purely scientific project is complicated by political tensions represented in the character of an unsympathetic Homeland Security officer overseeing the research, by the unusual, unnatural behavior of the wolves, and ultimately by the death of two researchers in what appears to be wolf attacks.

Charlotte McConaghy's standalone novel *Once There Were Wolves*[5] (2021) is set mainly in the Cairngorms National Park in the Scottish Highlands and features an amateur sleuth Inti Flynn, a biologist by profession, who arrives in the national park as part of a project to rewild Scotland by reintroducing wolves into the Cairngorms Mountains. She brings along her identical twin Aggie, deeply traumatized from an abusive marriage, in the hope that the natural setting would promote her healing. When a local farmer is found dead in what Inti fears local people would take for a wolf kill, she starts to investigate on her own to protect the wolves.

In the novels, the motif of a murder masked or perceived as an animal kill appears. Danielsson maintains that when a detective manages to prove that a human murderer killed the victim and thus exonerates the wrongly accused animal, the detective "is established as a protector of animals and supporter of environmental and animal preservation ethics" (Danielsson 105), a stance that strengthens the purposefully environmental appeal of the selected works.[6] Thus, despite their variations, the novels share many features which allow for a meaningful comparison to be drawn. They also reflect the recent leading trend in the genre – the predominance of female authors and protagonists – as well as confirm Jim Dwyer's claim in *Where the Wild Books Are* that a standard police procedural, as the most formulaic subgenre of crime fiction, is rarely employed for addressing environmental issues (Dwyer 165).

2   The Genre, Nature and Consolation

It has been established that among the appeals of crime fiction is its ability to provide readers with a sense of comfort, or, as Christine Ann Evans puts

5   The novel also received high critical acclaim, winning the Indie Book Award for Fiction in 2022 and a Nautilus Gold Award.
6   Danielsson persuasively shows the shortcomings of Barr's revisiting this plot device in *Blood Lure*, where it is given a problematic twist and a solution.

it, the "characteristic reception" of crime fiction resides in the "consolatory pleasure" it affords the reader (159). As early as the 1940s, W. H. Auden (1948) famously compared the effects of crime fiction on its readers to those of Greek tragedy on its spectators. More recently, Janice Law Trecker (2013) has asserted that crime fiction can provide consolation by showcasing the power of human intellect to solve mysteries (337). It fosters a sense of empowerment in "ordinary mortals" by suggesting that they can provide answers and achieve justice (337). The genre's denouement offers closure, a sense of control, and a form of consolation in the face of the uncertainties and complexities of our often chaotic world.[7] Furthermore, crime fiction presents a world where crime is identifiable, solvable, and explicable, asserting that justice is being pursued and maintained (see Raskin 1992; Knight 2003; Phillips 2016; Brewster 2017). So paradoxically, although concerned with crime among humans, the genre often ends on a hopeful note concerning human relationships.

I also argue that crime fiction can incorporate the theme of human–nature relationships and, therefore, of consolation in other ways. As evidenced in the selected novels, a change in the approach to the setting is a necessary prerequisite for addressing the human relationship with the more-than-human[8] world meaningfully. The novels no longer employ their respective settings as mere backdrops, providing traces and clues, but instead present them as authentically drawn complex environments. Thus, in the novels, nature (or wilderness) is not anthropomorphized, depicted as a stylized place with various symbolic meanings, human projections, and interpretations, but is approached as an environment as characterized in Lawrence Buell's *The Environmental Imagination*. It is viewed as a gapless space with a mutually interconnected network of live organisms and processes that humans participate in and depend on. All senses take part in the perception of the natural environment; the narrator (or character) is immersed in it, surrounded by it rather than separated from it as an observer (Buell 7). The perception of nature is a synesthetic or multisensory experience in which visual input, smells, sounds, tactile perceptions, and occasionally even tastes participate.

Accurate descriptions of natural environments prevail in Barr's and McConaghy's books. Despite the primary focus of the crime narrative on the action, attention is also paid to the geomorphological and climatic features of

---

7   This can be well documented by the tremendous increase in the popularity of crime fiction during the Covid pandemic when it offered the restoration of order, "that sense of resolution" (see Price 2020 and Wood 2020 as qtd. in Lucyna Krawczyk-Żywko 2021).

8   Coined by David Abram in his work *The Spell of the Sensuous* (1996), the term has been widely used in ecocriticism.

the respective environments, to their fauna and flora. For example, throughout her series Nevada Barr relates the formation of the cave system in New Mexico (in *Blind Descent*), the importance of wildfires for the regeneration of primal forests (in *Firestorm*), the effects of avalanches and mudslides on vegetation, and the habits of bears (in *Blood Lure*), of loggerhead turtles (in *Endangered Species*), and of wolves and moose (in *Winter Study*).

Charlotte McConaghy's *Once There Were Wolves* explains the role of apex predators in sustaining the biodiversity of an ecosystem. By doing so, the novel places the motivation behind the reintroduction of wolves into their former habitat within an ecological framework. This contrasts with the promotion of the 1995 project to reintroduce wolves to Yellowstone National Park, which, as Alison Byerly points out, often authenticated the wolf's value by comparing it to great works of art, emphasizing its symbolic value rather than its environmental significance (Byerly 58). The conclusion Danielsson makes about Barr's series is equally valid for McConaghy's novel even though wolves are not (yet) reintroduced[9] in the Scottish Highlands: "Its claim to verisimilitude extends to other aspects than descriptions of nature [and] include[s] verifiable particulars of park geography, history, policy and management, and data about the typical park species" (Danielsson 109). These particulars are always brief, usually unobtrusively incorporated into the narrative structure, and scientifically accurate.

The descriptions of landscape and weather are generally not to highlight or echo a character's mood or set the story's atmosphere but to convey information about the place. Sometimes, they are related matter-of-factly, for example when a biologist, upon finding a gnawed skeleton of a moose, explains that the fact that even its antlers have been nibbled indicates how hard the current winter is on the wolf population: "There's little nutritional value in an antler. Eating it is the animal world's equivalent of boiling shoe leather for supper" (Barr, *Winter Study* 126–127). Sometimes, information is imparted in a more poetic tone: "The day was painfully bright and clear as it can only be in the north where every particle of moisture is frozen from the air and the sun moves low in the south [...] crystalline amber light honed the edges of the world till shadows of pines [...] were as sharp and black as fangs drawn by children" (Barr, *Winter Study* 4). Occasionally, it is presented in a light, humorous tone, as this example about the arid country fauna's circadian rhythms: "With morning's peace came the animals: those just coming on diurnal duty, those

---

9 In an interview with Julie Carrick Dalton, McConaghy named the Yellowstone Wolf Project (the reintroducing of wolves into the park in 1995 and researching them ever since) as her source of inspiration and study material.

going off nocturnal shift, and the crepuscular crew with a split shift framing the day" (Barr, *Ill Wind* 31). The information does not pertain to the crime narrative but is introduced merely to highlight the diversity of biorhythms. Such is the case with most information about wildlife, climate, and geography inserted in the crime fiction under consideration.[10] Sometimes, the information explicitly underlines the works' environmental advocacy, for example in this brief description: "The trap was an ugly thing, a reminder of the metal rampage humanity with its mines and forges and industry had loosed on the world" (Barr, *Winter Study* 159).

Interestingly, inserting so much scientific information does not hinder the crime plot. On the contrary, it adds a welcome enlivening to the genre formula and boosts the authenticity of the setting and the reliability of the protagonist. In their incorporation of scientific data, the crime novels resemble the new nature writing in which, as Robert Macfarlane maintains, a "tonal mix of the poetic and the scientific and analytical" (166) is typical. However, unlike the new nature writing characterized by its generic indeterminacy, the novels are rather easy to categorize with regard to their genre, as shown above.

In terms of the crime genre, the most effective is incorporating scientific information, which is shown as vital for the investigation, i.e., when it eventually turns into a clue. In this example, the usage of the Latin name for the wolf and French for its eating habits aims at strengthening the objectivity of scientific discourse: "*Canis lupus* were designed to eat *en famille* and efficiently; an adult wolf's jaws exerted fifteen pounds per square inch, about twice that of a German shepherd and five times that of a human being. A mature wolf could gnaw through a femur in six or seven bites" (Barr, *Winter Study* 263, italics in the original). The information later helps determine whether a victim was killed by wolves or murdered by a human being trying to mask the murder as a wolf kill.

Apart from mixing elements of scientific discourse into crime narrative, the works also share the motif of escaping into nature, of looking to the wilderness for spiritual regeneration. There are good exemplifications of the so-called "wilderness therapy," a value ascribed to nature since the mid-twentieth century, when, as Roderick Frazier Nash explains, wilderness became viewed as a place to regain one's mental health, to slow down and relax, to find relief from the noise and stress of urban life (266).

---

10   I have addressed this aspect of recent crime fiction in more depth in Šárka Bubíková and Olga Roebuck, *The Place It Was Done. Location and Community in Contemporary American and British Crime Fiction* (McFarland, 2023).

The motivation for Barr's Anna Pigeon to start working with the National Park Service is her need to cope with the trauma of her husband's death in a car accident. She moves to her first national park service assignment in Guadalupe Mountains National Park in Texas directly from Hell's Kitchen in New York City and confirms that "[t]he beauty of the Chihuahuan Desert had been smoothing the wrinkles from [her] mind" (Barr, *Track of the Cat* 85). The comfort she finds in nature is described on many occasions, for example in the series's opening:

> In the pines, pygmy nuthatches and mule deer for companions, the breezes blew away thought and Anna achieved the peace she'd come to depend on the wilderness for. In place of doubt and suspicion came yellow butterflies feeding incongruously at fox dung. Unself-conscious as a bird, she whistled, lending human notes to the sweet cacophony of the woods. (119)

Throughout the series, Anna repeatedly states that she finds solace away from people, in nature: "The woods [...] were where she hid from the monsters of the populated world" (Barr, *Winter Study* 286). She even explicitly calls hiking alone in the wilderness "a medicine" (Barr, *A Superior Death* 153). Similarly, the belief in the therapeutic propensity of wild nature prompts McConaghy's protagonist Inti to bring her deeply troubled twin sister along to the Cairngorms Mountains.

Overall, McConaghy's novel evokes human kinship or unity with nature on many occasions. As a child, Inti believed that the trees of the forest surrounding her father's cabin were her "family" (15) and that their roots were her own. As an adult, she still occasionally desires to melt into the natural environment. Frustrated with her fellow humans, she escapes into a remote valley and, resting on the bank of a loch, she admits: "I want never to leave this place. [...] The aloneness is exquisite; it is calm" (McConaghy 73). The unity with the natural world is symbolized by *Werner's Nomenclature of Colours*, a book Inti's father (a former logger turned naturalist) cherishes explicitly for its unifying power: "It connects things [...] makes us part of nature" (McConaghy 16). Inti enjoys the wonder that the same hue can be found on human skin, stones, flowers, and bird feathers. While color has been used as a dividing factor among people – Frederick Douglass and W. E. B. Du Bois's usage of the term "the color line" easily springs to mind here – McConaghy, on the contrary, uses colors to instead highlight similarities, extending them to include both human and more-than-human worlds.

The sense of a special connection between her and other species, or what Chris Rutledge in his 2021 review for the *Washington Independent* calls "sensual connection to nature," is emphasized in Inti's relationship with wolves. Occasionally, she even seems to want to join the wolf pack. The events in her life parallel those of the wolves. Like the wolf pack, Inti, too, arrives in an unknown territory in a foreign country and tries to carve a space for herself there. She must learn to understand her fellow humans as well as the landscape and climate of the place. The wolves' task is to rewild the highlands and to help heal the national park's ecosystem by reducing the environmental damage of deer overpopulation. Similarly, Inti is attempting to help her twin sister heal from the trauma of an abusive and violent marriage. Like the wolves, she painfully clashes with local farmers and seems unable to ignite their enthusiasm for the rewilding project. Eventually, just like the wolves, she establishes a more or less working relationship with her new environment and settles down. The wolves mate, and Inti becomes involved with a local man. While the rewilding team observes the wolves' denning, Inti becomes aware of her pregnancy and the life growing within, yet she is highly reluctant to embrace the fact. When she is forced to put down a female wolf who keeps attacking livestock, she almost dies in the process. However, by killing her, Inti seems to have killed her own wild anger, and she is finally able to accept her new role as a mother.

Inti suffers from a rare developmental cognitive condition, the so-called mirror-touch synesthesia, which makes her feel the sensations she sees. In the narrative, the condition serves as an embodiment of empathy and universal kinship. Due to it, Inti is able to truly walk in somebody else's shoes, to experience somebody else's pleasure and pain. Even Inti's name, although of Incan origin and referring to a solar deity, evokes intimacy in English. Inti has a unique connection to other people and non-human beings through her condition. When she, for example, slips a rope around a horse's neck, she feels the texture and pressure of the rope on her own neck (McConaghy 36). If she sees somebody's face slapped, she feels the pain of the impact on her face, when she witnesses somebody losing their balance, she herself stumbles (McConaghy 183), etc. Thanks to her unusual perceptions, Inti transcends the boundaries of her own body and self and can truly become a selfless person. Nevertheless, the heightened sense of empathy is frequently a burden, making Inti vulnerable and exposed. Apart from the synesthesia-caused closeness to others, Inti has a very deep bond with her twin sister, in which the clearly defined boundaries of individual life often dissolve. By foregrounding the variety of intimacies, the narrative thus reflects on the interconnectedness of all forms of life, calling for empathy and responsibility in human dealings with each other and the more-than-human world.

Both authors refrain from romanticizing nature. They show it as indifferent to human strivings, often dangerous when people underestimate it. Barr's protagonist Anna poignantly observes: "Nature still retained the power to erase human lives as easily as she did the prints of their shoes" (Barr, *Winter Study* 125). Inti, although an experienced biologist, is an outsider to the Scottish Highlands and misjudges the harshness of its winter storms. Consequently, she almost dies of exposure when she pursues a wolf and gets caught birthing on a freezing night in the Cairngorms.

3   National Parks as a Setting

The selection of national parks as a setting means that both authors, in one way or another, engage in an ongoing conversation about the definitions, meanings, and value of wilderness and nature conservation. As I have summed up elsewhere (Bubíková 2021, 95–108), there is a rich and varied body of scholarship on (the cultural and literary meaning of) wilderness that shows how ambiguous and complex the concept is. As such, it has become a subject of vigorous criticism, especially from environmental and postcolonial positions. It is mainly the former that informs the selected narratives.

In her insightful essay on the picturesque aesthetics informing the conception of the national park system, Alison Byerly shows the consequences of deriving the notions of wilderness and natural beauty from artistic concepts. She argues that this "aestheticization" turned nature into "a legitimate object of artistic consumption" (53) and national parks into providers of "a mythologized image" (58) of wilderness, often treated as a "picturesque commodity" (59). Byerly's critical concerns variously resonate in the novels.

By setting her narrative in the Cairngorms National Park, McConaghy implicitly questions the idea of national parks as picturesque wilderness reservoirs and emphasizes long-lasting human participation in the forming of an eco-system instead. Nowadays, the park is a permanent home to almost twenty thousand people who live there in numerous towns and villages. Throughout history, it has served as a refuge during religious wars, as a place for illicit distilling, and, on a much larger scale, for agriculture and intensive sheep grazing. It has gone through deforestation since the Middle Ages and extensive afforestation for timber production in the last century (Oosthoek 33–50). Aware of the illusory nature of the traditional understanding of wilderness, McConaghy hardly uses the term and speaks of wildness instead. The concept of wildness applies to organisms and ecosystems, not just places, and, as Eric S. Higgs explains, "locates the power of meaning in process rather than place"

(503). In that way, McConaghy does not highlight the illusionary quality of an undeveloped natural habitat but its ability to renew itself. She also echoes the recent discussion on the vitality of the concept of wildness,[11] applying it simultaneously to wildlife and to people.

Therefore, although the novels are situated in officially established national parks, they also draw attention to the fact that the areas designated as national parks vary greatly. Because local conditions regarding the degrees of human interaction with a particular environment and nature conservation management can differ substantially, the International Union for Conservation of Nature (IUCN) developed a system of six protected area categories. What is designated under various local legislations as a national park is not necessarily the equivalent of the IUCN "national park" category, or, as Adrian Phillips and Jeremy Harrison laconically conclude, "the term 'national park' means different things in different countries" (Phillips and Harrison 29). The Cairngorms National Park, for example, is not classified as a (category two) national park according to the IUCN but as an area in which people have interacted with the landscape for many years and which is managed to sustain the habitats and landscape that have resulted from this interaction (category five) (Dudley 2).

However, differences can be found even within one country, as Barr's series effectively shows. Since her protagonist's job takes her from park to park, she often not only compares natural conditions but also the variety of areas comprised under the National Park Service system. She distinguishes, for example, the pristine wilderness of Glacier National Park, the restored wilderness of Isle Royale National Park, the protected archaeological site of the Mesa Verde National Park, the protected recreational area of Glen Canyon, etc.

Although Barr also acknowledges human participation in the formation of environments, she is more or less silent about the presence of Indigenous peoples in areas nowadays designated as national parks. One rather obvious exception is the Mesa Verde National Park, where the preservation of Indigenous cultural heritage is a significant part of the park's mission, and therefore, the novel set there naturally foregrounds it, for example, in the description of the Spruce Canyon. The canyon is co-created by natural forces as well as human activity, layered for centuries:

---

11   See for example John Hausdoerffer, "The *Akiing* Ethic"; also Kim Ward, "For Wilderness or Wildness? Decolonising Rewilding," *Rewilding* (2019): 34–54; Lawrence J. Cookson, "A Definition for Wildness," *Ecopsychology* (2011): 187–193. doi 10.1089/eco.2011.0028; Patrick Ram Kelly, *The Enduring Importance of Wildness: Shepherding Wilderness Through the Anthropocene* (2018).

There the mesa fell away in staggered steps of fawn-colored sandstone, before a sheer drop to the wooded ground below. Like many canyons cut into the mesa, Spruce was small. [...] Each [canyon] had its own dwellings, long since abandoned by their owners and bleached back to the color of the earth. Since Mesa Verde's cliffs had first been inhabited the Anasazi, the Utes, the Navajo, cowboys, hunters, and tourists had all tramped the trails. (Barr, *Ill Wind* 151)

Barr briefly mentions the Indigenous history of Isle Royal but hardly refers to it in her novels set in other national parks, although Native American presence has been evidenced in nearly all of them.[12] Often, Native Americans were forcefully removed or fatally cut off from their natural livelihood once a national park was established, and even though this part of the national park system's history was unheralded, it is now openly acknowledged and discussed even on the national park service's web pages. In this respect, Barr's silence is reminiscent of an outdated governmental attitude to Native American presence in the "wilderness" of national parks.[13]

Thus, both novels implicitly call attention to the complex issues surrounding national parks' establishment, management, and purpose by employing the perspective of a professional environmentalist. Barr's protagonist often draws attention to visitors breaking nature protection rules in national parks, endangering the delicate balance of the ecosystems, and she occasionally challenges certain park management procedures or decisions (for example, the plan to open Isle Royal NP to the public for winter recreation in *Winter Study*). On the whole, though, she still represents the National Park Service authority and policies, narrating from within the system of governmental nature protection. Barr's perspective is that of a law enforcement officer or, in the words of Alison Byerly, of a "protector of valuable commodities" (59). Barr's narratives

---

12   For example, the Guadalupe Mountains National Park (the setting of Barr's *Track of the Cat*) used to be a sanctuary to the Mescalero Apaches, Lassen Volcanic National Park (where Barr's *Firestorm* is set) was a summer home to Atsugewi, Yana, Yahi, and Maidu people, the Miwok people have lived in the Yosemite national park (the setting of Barr's *High Country*), etc.

13   In Barr's fiction, there is not only a conspicuous silence regarding the presence of Native Americans but also regarding the issue of environmental injustice, as Ruth Hawthorn points out in her short note on Barr's *Burn*. According to Hawthorn, Barr avoids social and racial aspects of the Katrina disaster, offering only "brief and uncritical mentions of the city's uneven recovery" (298). Both these omissions can be considered a significant oversight, given the importance of the treatment of indigenous populations and environmental injustice in the broader environmental discourse.

thus often operate on the dichotomy of the park (i.e., nature) and its visitors (i.e., people).

McConaghy's Inti has a more ambiguous position because she is a scientist, not a law enforcement officer. She is an outsider to the Cairngorms Mountains, even to Scotland, because she is Australian and spent part of her childhood with her father in Canada. She also spent part of her adulthood in Alaska researching wolves in Denali National Park. In the novel's opening, she is portrayed as having little patience with the local farmers' fear of wolves and apprehension of the project's impact on their sheep farms and traditional livelihood. Inti manifests more empathy and concern for the wolves than humans, and her co-workers have difficulty preventing her from further antagonizing local people. It is only through her gradual involvement in the local community and budding relationship with Scottish police chief Duncan that she becomes more empathetic as well as aware that the attitude towards the wolves is not unanimous and that some local people are already involved in a variety of small-scale rewilding efforts. McConaghy thus emphasizes the communal character of nature preservation – it cannot be left up only to enthusiastic individuals nor just to governmental authorities, but everyone can, and indeed must, contribute because, as Nathan J. Bennett et al. propose, "environmental stewardship actions can be taken at diverse scales, from local to global efforts, and in both rural and urban contexts" (597). The individual contribution to environmental stewardship does not mean just obeying nature protection rules when visiting a national park but actively participating in nature protection and rewilding, especially in one's own place.

Inti comments on the re-wilding process: "And when you open your heart to rewilding a landscape, the truth is, you're opening your heart to rewilding yourself" (189). In her case, this means opening her heart for regeneration, allowing human trust and love its proper place in her life. It is the trust in non-human creatures that eventually helps her regain her trust in humanity. In this way, McConaghy confirms David Kidner's claim in "Culture and the Unconscious in Environmental Ethics" (1998), that "in a healthy culture, the wildness/otherness within us resonates with parallel aspects of the rest of the natural world" (72). In other words, the process of rewilding, i.e., accepting that humanity belongs to "the symbolic community of nature," and furthering "the resonance between our own wildness and that in the rest of the world," will eventually "reinforce the wholeness of nature" and help in healing "the splits between what is 'wild' and what is 'civilized,'" the splits Kidner sees as results of our technocratic perception of the world (74).

In the denouements of both novels, the culprits are removed, and the protagonists realize with a degree of relief that not only their scientific understanding of wolves based on years of research and observations was accurate and they are not man-killers, but that only certain human beings rather than the whole of humankind are violent and abusive.

4 Predators and Violence

Apex predators have long been recognized as tourist magnets for many national parks. Unsurprisingly, therefore, wolves feature in Barr's *Winter Study*, and they are central to McConaghy's novel. Wolves hold a prominent position in the Western literary and cultural imagination. As Peter Hollindale summarizes, "no wild creature is more self-destructively proficient than the wolf at creating irrational fear in human beings" (Hollindale 97), and he not only lists traditional tales featuring wolves but also English idioms about them, all of which "teach to fear and demonize the wolf" (98). Both novels contain observations of wolves made directly either by the protagonists or by scientists in their respective teams, and in this way, the descriptions are given the semblance of reliable scientifically collected data. Human preconceptions and fears about the species are contrasted with real, observable wolf behavior. In *Winter Study*, a researcher's body is discovered and when a Homeland Security officer suggests a wolf pack hunted her down, a scientist contradicts him: "If it was wolves – natural, pure wolves – it's the first time in recorded history it's happened in America" (215). Similarly, the biologists in Inti's team try to calm local farmers by citing statistics[14] and describing the wolf as "a shy, family-oriented, *gentle* creature," emphasizing that people "should never have been taught to fear [wolves]" (McConaghy 26, italics in the original). Barr's protagonist likewise notes that "[w]olves are a private people, a quiet, watchful people" (Barr, *Winter Study* 107).

Winter Study refers to an existing scientific research project[15] concerning wolves to which the protagonist, Anna, is invited. She describes her first

---

14   Such as the conclusion of John D. C. Linnell, Ekaterina Kovtun, and Ive Rouart, *Wolf Attacks on Humans: An Update for 2002–2020* (Trondheim: Norwegian Institute for Nature Research, 2021), which says: "In Europe and North America we only found evidence for 12 attacks (with 14 victims) of which two (both in North America) were fatal, across a period of 18 years. Considering that there are close to 60,000 wolves in North America and 15,000 in Europe, all sharing space with hundreds of millions of people, it is apparent that the risks associated with a wolf attack are above zero, but far too low to calculate" (3).

15   The wolves and moose research project is "the longest continuous study of any predator-prey system in the world." For more information, see the project on the Isle Royale NP web page: https://isleroyalewolf.org/overview/overview/at_a_glance.html or on the Michigan

sighting of the wolves moving single file with their "heads low, [...] long legs and big paws carrying them effortlessly over the patchy snow [...], the moon catching their fur until they were frosted with silver" (39–40). She calls the sight "pure magic" (40). Yet she is well aware that "in the Western world's collective unconscious, wolves symbolized hunger, danger, vicious cunning and cold-blooded slaughter" (62). To show how deep such notions and fears are lodged, even Anna must occasionally remind herself to differentiate between real wolves and mythological monsters when she finds herself in the middle of the freezing, isolated winter landscape under the constant threat of an unknown murderer whose crimes appear as wolf attacks.

In *Once There Were Wolves*, Inti describes wolves as "subtly powerful, endlessly patient, and more beautiful than anything [she has] seen" (McConaghy 178). She is thus stunned by the intensity of the hatred felt by some farmers toward wolves as well as by the thoroughness with which wolves had been hunted down and completely erased from Scotland. She is shocked when a farmer kills a wolf just because he spotted it, and she is abhorred when a local youth mutilates a wolf to mock the rewilding efforts. She expresses her anger at hunters who kill wolves just for the sake of killing, to feel "power over another creature" (147) and have a trophy to document it. Anna is equally upset that "these intelligent and phenomenally complex animals could be hunted down and butchered so that some fool could have the pelt and head for a hearth rug" (Barr, *Winter Study* 93). The way wolves are treated in the narratives again underlines the novels' environmental concerns and contributes to raising awareness about the predator. As Kateřina Kovářová says: "Describing the predator as beautiful and vulnerable is crucial for changing the way in which the wolf is imagined" (Kovářová 64).

In their reflections on the nature of crime, the novels contrast human violence with violent events in nature. In both novels, a wolf is found dead, apparently killed in the packs' fight over territory. The protagonists feel sadness yet no anger, emphasizing that this is a natural behavior of wolves. The existence of wolf pack wars is acknowledged in both McConaghy's and Barr's novels. Anna explains how they differ from human wars: when the packs clashed, "it was hit-and-run, not the full-scale slaughter humans had perfected" (Barr, *Winter Study* 42). Likewise, while acknowledging the fatal results of animal predation or fights between animals, they distinguish it from human evil on the basis of the absence of wickedness. In the words of Anna: "The natural world lived and died by tooth and claw, without malice and without regret" (Barr, *High Country*

Technological University web: https://www.mtu.edu/news/2023/06/isle-royale-winter
-study-wolf-count-rises-slightly-moose-population-drops.html.

58). However, in "people killing people [...] always there was evil" (Barr, *Blood Lure* 70). It is malice and intentionality that differentiates between a predator killing its prey and a murderer. Also, unlike such natural incidents, crime is not part of the concept of wildness because it is not followed by regeneration, only by pain and loss. The detailed scene of a wolf pack hunting and killing a moose is presented as a natural, unavoidable, even dignified event, which Anna sees as "a beautiful dance of life and death" (Barr, *Winter Study* 51). It is necessary for the survival of the wolves and beneficial for the ecosystem in the long run. On the other hand, all the murders in the novels are seen as tragedies, unnecessary wastes of human lives with no benefit to the environment.

To emphasize how misguided the human fear of wolves as killing monsters is, the novels contrast canine predation with human violence towards wildlife. In both, a human kills a wolf just to make a point, to show power and control. The animal is not killed in (self)protection nor for food, only out of malice. Natural predation is also placed in contrast to violence among humans, of which domestic violence stands out as the most horrendous kind. In *Once There Were Wolves*, domestic violence is among the major themes of the novel, and it also forms an important motif in Barr's *Winter Study*. Barr's culprit is revealed as a cold-blooded murderer and a rapist, as well as a survivor of child abuse. Inti explicitly points out that it is not wolves who are beasts but humans: "If you truly think wolves are the blood spillers, then you are blind. *We* do that. We are the people killers, the children killers. *We're* the monsters" (McConaghy 26, italics in the original). And Anna likewise concludes: "Man gave the wolf all the dark bits of himself, then vilified the wolf" (Barr, *Winter Study* 327).

When both protagonists reveal the assumed wolf kill to be murder, they exonerate the wrongly suspected animals in agreement with the novels' environmental focus and biological verisimilitude, just as Danielsson (2021) points out in her analysis of Barr's *Blood Lure*. The reader is thus offered the typical consolation of a mystery solved, but this comfort is given an environmental twist when it turns out that the wolf is indeed not a murderer. Considering how both narratives manipulate their readers to empathize with the animals, such a resolution strikes a note of hope concerning human-wolf cohabitation, even if McConaghy's denouement is more ambiguous – the truth about the murder is revealed to the readers, but not the characters in the novel. Still, the novel provides the relief that wolves are not the monster killers people assume them to be.

Both novels, therefore, resonate with Tokarski's ideas expressed in "Consolations of Environmental Philosophy" (2021), which are elucidated in this volume's introduction. He suggests that shifting our perspective from a human-centered approach to the non-human world and understanding our interconnectedness with the natural world is an effective strategy for fostering a more compassionate and harmonious relationship with other species (455). At the same time, human fear, anxiety, or distress experienced in encounters with non-human creatures such as wolves can be greatly reduced (459). Both novels attempt to present such a shift in perspective through the perceptions and attitudes of their respective protagonists.

5 Conclusion

The analysis has shown that contemporary crime fiction addresses environmental issues in a nuanced manner, beyond merely making the crime under investigation an environmental one. These works resemble new nature writing by combining scientific and poetic discourses within their narrative structures. Unlike new nature writing, however, they are clearly classified as crime novels. While they incorporate the motif of escape into nature, they problematize it by making nature the setting for crime and investigation.

In these novels, synesthesia is used in descriptions of nature and wildlife as a multisensory perception of the natural environment. McConaghy's novel takes it further by using Inti's condition, mirror-touch synesthesia, to symbolize the interconnectedness of all living creatures and the human capacity for empathy. This capacity is contrasted with violent crimes.

Set in national parks, the novels draw attention to issues of national park classification, establishment, and management. From the perspective of a professional environmentalist, they challenge preconceptions about wilderness and the artificiality of the nature-civilization dichotomy. They critique the concept of wilderness as a self-willed place, highlighting human participation in the creation of environments, even in remote areas such as the Scottish Highlands and Isle Royal archipelago. Rather than viewing national parks solely as places for recreation and education, they present them as places of wildness (i.e., of regeneration and renewal) and as a laboratory for scientific experiments.

Predators such as wolves, hallmarks of wilderness, are central to the narrative. The novels aim to present wolves with verisimilitude, realistically rendering their characteristic features, behavioral patterns, and irreplaceability in sustaining biodiversity. Wolves are portrayed as magnificent creatures who can also become victims of human actions. By using both the symbolic figure of a predator and scientifically accurate descriptions, the novels address the misrepresentation of wolves as blood-thirsty human killers and contrast natural predation for sustenance with human-induced violence, either towards the environment or other humans. They foreground domestic violence as a particularly abhorrent form of predation humans are capable of.

Following the detective fiction formula, the novels use the motif of a murder masked as a predator kill, with the detective's responsibility being to exonerate the animal and identify the real killer, the human. This duality portrays humans as capable of empathetic kinship with wildlife and responsible environmental stewardship, while also being the true dangerous killers. The novels provide an additional consolation by confirming that wolves are not the blood-thirsty monsters of folklore tradition and that peaceful human–wolf cohabitation is possible if people abandon a human-centered perspective and learn to understand and respect non-human creatures in their own right.

Bibliography

Abram, David. *The Spell of the Sensuous*. Vintage, 1996.

Ashman, Nathan, editor. *The Routledge Handbook of Crime Fiction and Ecology*. Routledge, 2024.

Auden, W. H. "The Guilty Vicarage: Notes on the Detective Story, by an Addict." *Harper's Magazine*, May 1948, https://harpers.org/archive/1948/05/the-guilty-vicarage/.

Barr, Nevada. *Track of the Cat*. 1993. Berkeley Books, 2003.

Barr, Nevada. *A Superior Death*. 1994. Berkeley Books, 2003.

Barr, Nevada. *Ill Wind*. 1995. Berkeley Books, 2004.

Barr, Nevada. *Winter Study*. 2008. Berkeley Books, 2009.

Bennett, Nathan J., et al. "Environmental Stewardship: A Conceptual Review and Analytical Framework." *Environmental Management*, vol. 61, 2018, pp. 597–614. doi: 10.1007/s00267-017-0993-2.

Bertens, Hans, and Theo D'haen. *Contemporary American Crime Fiction*. Palgrave McMillan, 2001.

Brewster, Liz. "Murder by the Book: Using Crime Fiction as a Bibliotherapeutic Resource." *Medical Humanities*, 43, 2017, pp. 62–67. doi: 10.1136/MEDHUM-2016-011069.

Bubíková, Šárka. "Wilderness in Dana Stabenow's and Nevada Barr's Crime Fiction Series." *Places and Spaces of Crime in Popular Imagination*, edited by Šárka Bubíková and Olga Roebuck, Jagiellonian University Press, 2021, pp. 95–108.

Buell, Lawrence. *The Environmental Imagination: Thoreau, Nature Writing, and the Formation of American Culture*. Harvard University Press, 1995.

Byerly, Alison. "The Uses of Landscape: The Picturesque Aesthetic and the National Park System." *The Ecocriticism Reader: Landmarks in Literary Ecology*, edited by Cheryll Glotfelty and Harold Fromm, University of Georgia Press, 1996, pp. 52–68.

Clark, Timothy. *The Cambridge Introduction to Literature and the Environment.* Cambridge University Press, 2011.

Dalton, Julie Carrick. "Six Questions for the Author of *Once There Were Wolves.*" *Orion Magazine*, 4 Aug. 2021, https://orionmagazine.org/2021/08/six-questions-for-the -author-once-there-were-wolves/.

Danielsson, Karin Molander. "Ecology, Capability and Companion Species: Conflicting Ethics in Nevada Barr's *Blood Lure.*" *Animals in Detective Fiction*, edited by Ruth Hawthorn and John Miller, Springer International Publishing, 2022, pp. 105–125.

Dudley, Nigel, editor. *Guidelines for Applying Protected Area Management Categories.* IUCN, 2008, https://portals.iucn.org/library/sites/library/files/documents/pag-021 .pdf.

Dwyer, Jim. *Where the Wild Books Are: A Field Guide to Ecofiction.* University of Nevada Press, 2010.

Evans, Christine Ann. "On the Valuation of Detective Fiction: A Study in the Ethics of Consolation." *Journal of Popular Culture*, vol. 28, no. 2, 1994, pp. 159–167. doi: 10.1111/j.0022-3840.1994.2802_159.x.

Hawthorn, Ruth. "Hurricane Katrina and 'The City That Care Forgot'." *The Routledge Handbook of Crime Fiction and Ecology*, edited by Nathan Ashman, Routledge, 2024, pp. 295–307.

Hausdoerffer, John. "The *Akiing* Ethic: Seeking Ancestral Wildness beyond Aldo Leopold's Wilderness." *Wildness: Relations of People and Place*, edited by Gavin Van Horn and John Hausdoerffer, University of Chicago Press, 2017, pp. 195–204.

Higgs, Eric S. "Restoration Goes Wild: A Reply to Throop and Purdom." *Restoration Ecology*, vol. 14, no. 4, 2006, pp. 500–503. doi: 10.1111/j.1526-100X.2006.00161.x.

Hollindale, Peter. "Why the Wolves Are Running." *The Lion and the Unicorn*, vol. 23, no. 1, 1999, pp. 97–115. *Project Muse.* doi: 10.1353/uni.1999.0008.

Kidner, David. "Culture and the Unconscious in Environmental Ethics." *Environmental Ethics*, vol. 20, no. 1, 1998, pp. 61–80.

Knight, Stephen. "The Golden Age." *The Cambridge Companion to Crime Fiction*, edited by Martin Priestman, Cambridge University Press, 2003, pp. 77–94.

Kovářová, Kateřina. "A Vulnerable Predator: The Wolf as a Symbol of the Natural Environment in the Works of Ernest Thompson Seton, Jack London and Cormac McCarthy." *Mediating Vulnerability: Comparative Approaches and Questions of Genre*, edited by Anneleen Masschelein, Florian Mussgnug and Jennifer Rushworth, UCL Press, 2021, pp. 52–68.

Krawczyk-Żywko, Lucyna. "The Comfort of Crime: The Appeal of Formulaic Fiction during the Pandemic." *Litteraria Copernicana*, vol. 39, no. 3, 2021, pp. 13–22. doi: 10.12775/LC.2021.022.

Linnell, John D. C., Ekaterina Kovtun, and Ive Rouart. *Wolf Attacks on Humans: An Update for 2002–2020.* Norwegian Institute for Nature Research, 2021.

McConaghy, Charlotte. *Once There Were Wolves*. 2021. Chatto & Windus, 2022.

Macfarlane, Robert. "Environment: New Words on the Wild." *Nature*, vol. 498, 2013, pp. 166–167. doi.org/10.1038/498166a.

Nash, Roderick Frazier. *Wilderness and the American Mind*. 1967. Fifth edition. Yale University Press, 2014.

Oosthoek, K. Jan. *Conquering the Highlands: A History of the Afforestation of the Scottish Uplands*. Australian National University Press, 2013.

Phillips, Adrian, and Jeremy Harrison. "International Standards in Establishing National Parks and Other Protected Areas." *The George Wright Forum*, vol. 14, no. 2, 1997, pp. 29–38.

Phillips, Bill. "Crime Fiction: A Global Phenomenon." *Journal of Literature & Librarianship*, vol. 5, no. 1, 2016, pp. 5–15. doi: 10.22492/IJL.5.1.01.

Puxan-Oliva, Marta. "Crime Fiction and the Environment." *The Routledge Companion to Crime Fiction*, edited by Janice Allan et al., Routledge, 2020, pp. 362–370.

Raskin, Richard. "The Pleasures and Politics of Detective Fiction." *Clues: A Journal of Detection*, vol. 13, 1992, pp. 71–113.

Rutledge, Chris. "[Review of] *Once There Were Wolves*." *Washington Independent*, 5 August 2021, https://www.washingtonindependentreviewofbooks.com/index.php/bookreview/once-there-were-wolves. Accessed 4 March 2024.

Tokarski, Mateusz. "Consolations of Environmental Philosophy." *Animals in Our Midst: The Challenges of Co-existing with Animals in the Anthropocene*, edited by Bernice Bovenkerk and Jozef Keulartz, Springer, 2021, pp. 445–467.

Trecker, Janice Law. "Wilkie Collins's Sleuths and the Consolations of Detection." *The Midwest Quarterly*, vol. 54, no. 4, 2013, pp. 337–351.

Walton, Samantha. "Studies in Green: Teaching Ecological Crime Fiction." *Teaching Crime Fiction*, edited by Charlotte Beyer, Palgrave McMillan, 2018, pp. 115–130.

Walton, Jo Lindsay, and Samantha Walton. "Introduction to Green Letters: Crime Fiction and Ecology." *Green Letters*, vol. 22, no. 1, 2018, pp. 2–6. doi: 10.1080/14688 417.2018.1484628.

Index

Abram, David   38, 39–42, 212*n*
Alaimo, Stacy   88, 89, 146
Alexander, Neal   122, 130–31, 141
animism   38, 39, 46, 50, 91
Anthropocene   1, 10, 15, 18, 85, 86, 88, 89,
   129, 130, 161, 163, 164–70, 176, 178, 179,
   181–84
anthropocentrism   89, 101, 144, 165, 168
anthropocentric   6, 62, 65, 89, 89*n*6, 94, 98,
   101, 107, 124, 129–30, 136, 146, 150, 209
   See disanthropocentric
anthropomorphism   15, 85, 89–92
attentiveness   18, 27, 34, 161, 166–70, 173, 175,
   179, 181–183
Auden, Wystan Hugh   53, 54, 62–69, 212
autobiography   12, 26, 28, 28*n*3, 107, 148

Bacigalupi, Paolo   85
   "The People of Sand and Slag"
   (2004)   92–94, 101–2
Baden-Powell, Robert   54–55, 58, 59, 64, 74,
   82
Barr, Nevada   209–25
   Winter Study (2008)   209–25
   Blood Lure (2010)   210, 211*n*6, 213, 223
Baszile, Natalie   188, 199–204
   Queen Sugar (2014)   199–203, 204
Bator, Joanna   144, 151–56
   "Tikkun Olam" (2022)   144, 151–56
Bennett, Jane *Vibrant Matter* (2010)   87
Berssenbrugge, Mei-mei   161
   Hello, the Roses (2013)   179–82
biodiversity   1, 162, 213, 224
biophilia   7–8
Black agrarianism   187–204
Boethius   3, 28
   De Consolatione Philosophiae (*The
   Consolation of Philosophy*) [524
   CE]   3–4, 28–9
Buell, Lawrence   1, 208, 212
Burnett, Elizabeth-Jane   16, 107, 108, 108*n*1,
   110, 110*n*5, 112–24
   The Grassling (2019)   107–24

Clark, Timothy   86, 88, 129, 131, 163, 209

Conrad, Joseph   25–36
   *The Mirror of the Sea: Memories and
   Impressions* (1906)   25–36
conservation   11, 13, 17, 19, 77, 109, 128, 135,
   189, 208, 217–18
consolation   2–6, 9, 16, 26, 30, 33–34, 36, 53,
   54, 57, 58, 61–9, 92–94, 97, 98, 107, 110,
   111, 119, 123, 129, 130, 134, 142, 144, 145,
   155, 163, 169, 182–3, 187, 189, 204, 208,
   211–12, 223, 225
crisis   1–15
   climate crisis   2, 10
   See ecological crisis
   environmental crisis   1, 2, 8, 12, 85, 86, 89

Deakin, Roger   128, 132–42
   Waterlog (1999)   132–42
Dimmock, Frederick Haydn   54, 59, 67–68
disanthropocentric   85, 86, 89, 90, 97, 101
Donato, Antonio   3
dualism   86–89
   mind–body dualism   115–16
Dungy, Camille T.   161
   Soil: the Story of a Black Mother's Garden
   (2023)   170, 173–76

ecocritical   1, 10, 13, 14, 38, 39, 42, 65, 86, 89,
   130, 210, 210*n*
ecocriticism   1, 13, 38
   material ecocriticism   87–88
ecology   130, 131, 168
   dark ecology   10
   ecological abundance   177
   ecological awareness   9, 179
   ecological crisis/breakdown/catastrophe/
   decline   9, 10, 13, 111, 161, 162–64, 182,
   183, 193, 209
   ecological framework   197, 213
   ecological mesh   97
   ecological politics   130
   ecological reality   172
   ecological stewards   197
   ecological texts   65
   ecological thought/thinking   9, 10
   ecological uncertainty   166, 183